LAB

Mark C. Fishman

LAB

Building a Home
for Scientists

Lars Müller Publishers

Love and thanks to Martha, Eric, and Sarah,
who often see more clearly than do I.

WHAT IS A LAB?

It is far more than a place for scientists to get their experiments done. It is their home for most of their waking hours. Can this home make their lives more enjoyable through design and aesthetics? Certainly. Can it make them more creative and effective? Who knows. Great discovery centers have been uninviting, cramped, dimly lit, and smelly. But a Dickensian sweatshop is neither necessary nor sufficient for scientific discovery.

So why not be beautiful, healthful, and comfortable?
It is more than being pretty. Somehow the structure should embody the greatness of science itself and should challenge the occupants to surpass the great discoveries of their forebears. It should remind them of the privileges of working in scientific discovery, of deciphering nature and contributing directly to the health of mankind. And it should celebrate more than the technological cutting edge. Like a monastery, it should invite contemplation and the search for truth.

On a daily basis, scientists look forward. So the lab must be designed to discover and to synthesize the chemical and biological substances to mimic, test, and modify life. Today the lab must accommodate tasks beyond the capabilities of an individual, connect to colleagues across time zones, and capture and interpret terabytes of new information. And it must be flexible in design to accommodate new discovery tools and approaches.

This book reflects some of the lessons I learned from helping to design new lab buildings in the United States, Switzerland, and China. All are part of the Novartis Institutes for BioMedical Research (NIBR). NIBR is the division of the Novartis Pharmaceutical Corporation dedicated to the discovery of new medicines and to their testing in early clinical trials. It is a global organization. I focus here primarily upon sites in Cambridge, Massachusetts, USA, and Shanghai, China. (In Basel, the site of the corporate headquarters for Novartis Pharmaceuticals, the new lab buildings were part of a broader campus project dedicated to many other activities beyond science. Several books have been written about this inspirational campus.)

Who do we build for in NIBR? For the dreamy soul dedicated to the discovery of fundamental biological principles, and the utilitarian chemist focused on the design of specific molecules; for the introvert and the extrovert; for those working at the wet bench and those programming computers and those planning clinical trials. Spaces must accommodate solo discovery and interdisciplinary teams, often working across the oceans.

How to fit such instincts and endeavors under one roof? I have no prescription, but here I outline some of the forces influencing us as we gave suggestions to the architects about the history of science, the scientists, the technologies, the furniture and equipment, and about how space is used (or not) today. We thought about indoors and out. We thought about the creative process and how it might be fostered. And we considered whether any attributes differed between the campuses we built in Shanghai and Cambridge.

Having lived for nearly three decades before NIBR at the Massachusetts General Hospital and Harvard, and having designed new laboratories there, I can say there is nothing unique to a laboratory for pharmaceutical discovery. As far as I can see, the principles and even most of the details are universal.

In the building design, I had the opportunity to work with and to learn from more than a dozen world-class architects, because each building was designed by a different firm, so we have a wonderful set of distinctive answers to similar questions. This book is neither a paean to each building nor a prescription for precisely what to do. It is quite personal about which elements I deemed important, but these were built upon many hours of consultation with my fellow scientists and with the architects.

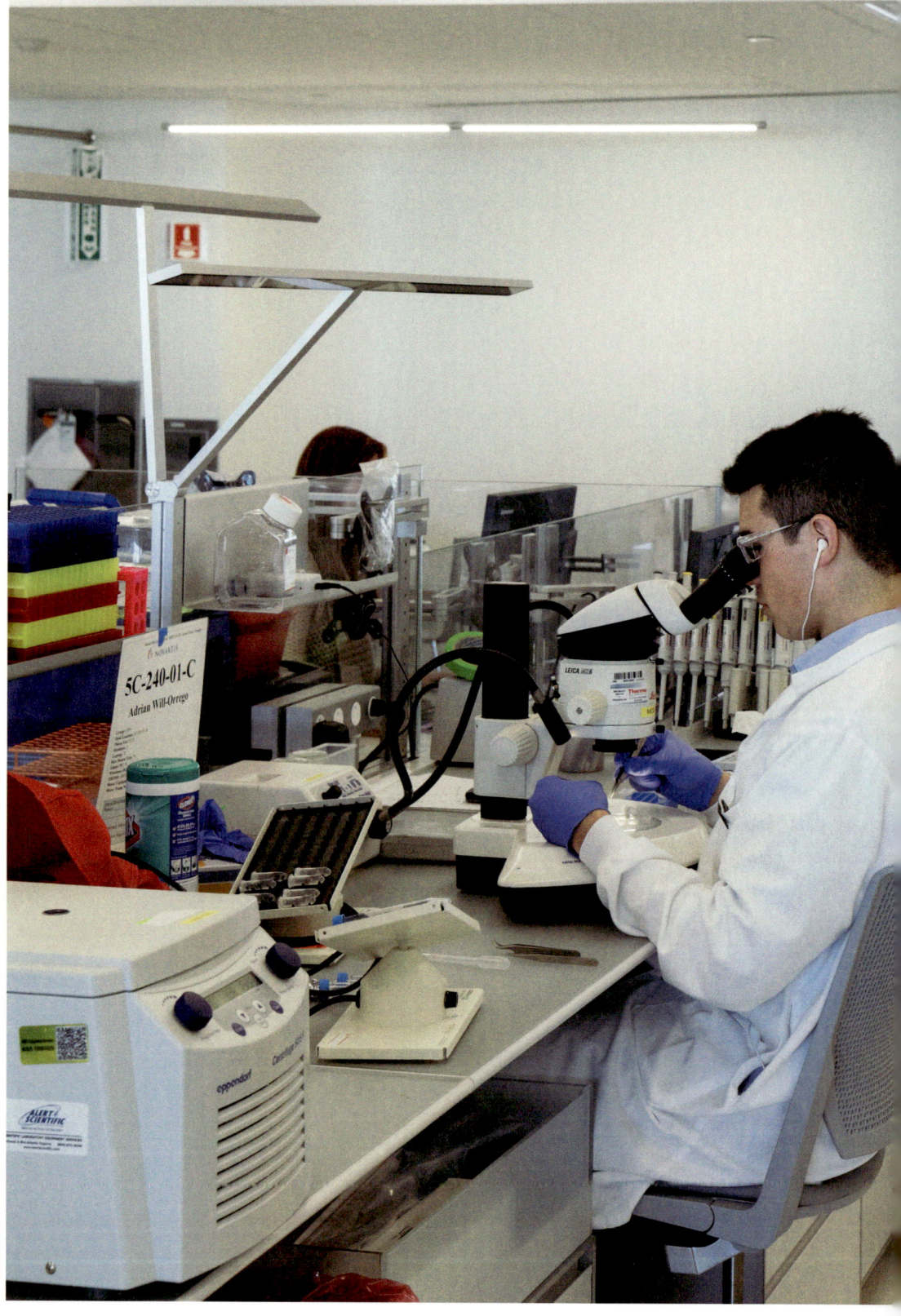

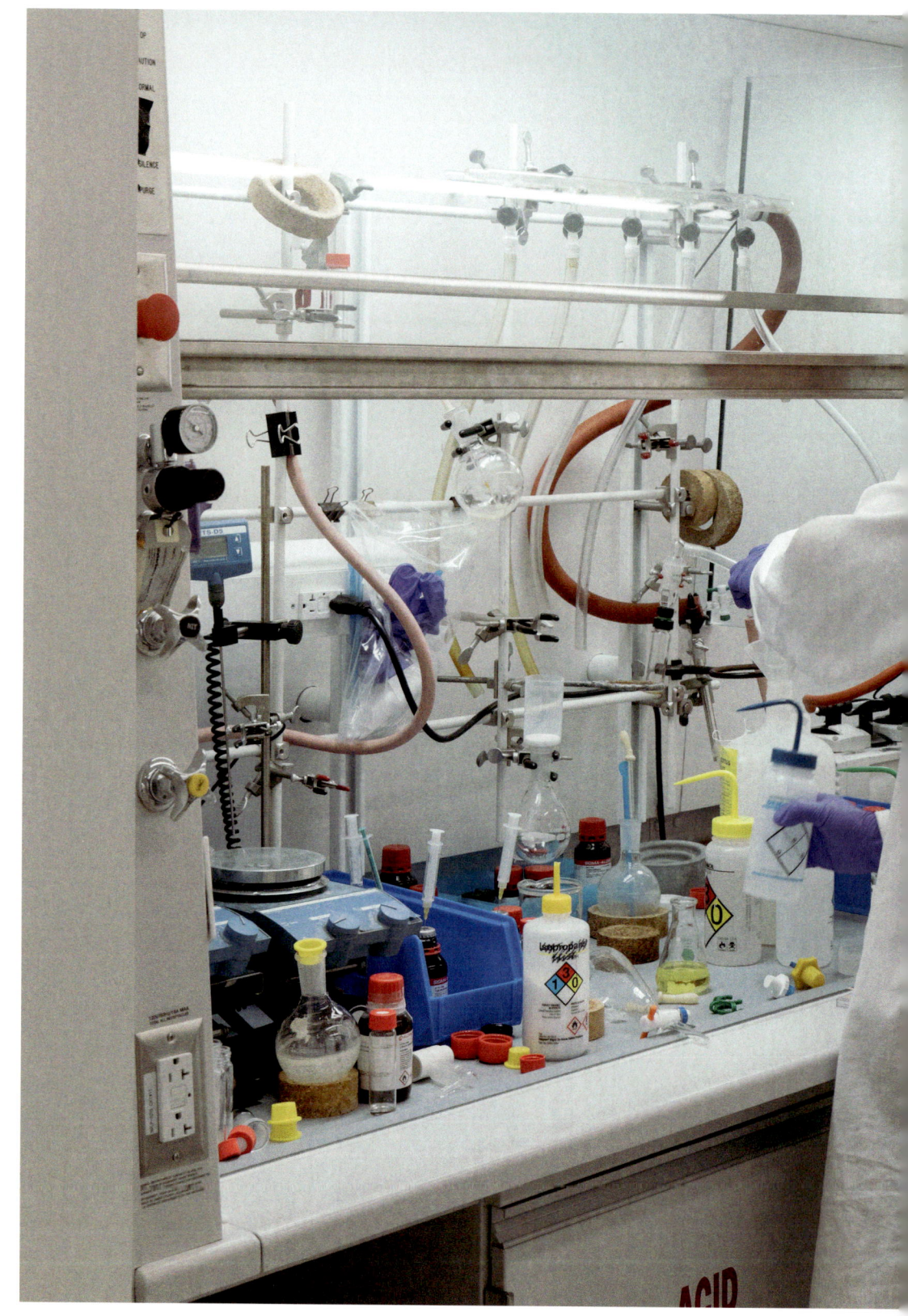

EXIT

EVOLUTION
CHANCE, BEAUTY, AND SIMPLICITY

NATURE:
CONSERVATIVE AND BEAUTIFUL

Can the environment enhance the creativity and success
of the scientist? Familiarity and dissonance, clarity
and confusion, quiet and noise affect each person
differently, and scientists hail from all over the world,
with different training and expectations. But they
do share a heritage of certain great ideas and unspoken
understandings.

A central theme of biology and medicine today is the
conservatism of nature, the use and reuse of systems and
structures throughout the animal kingdom. This fortu-
nate occurrence allows us to generalize from cells
in culture to organs in humans and from development
of the fruit fly to therapy for cancer. The origin of evidence
for this continuity of nature is the theory of evolution,
and its modern incarnation is genetics. The same
few conserved molecular pathways are used in slightly
different formulations for all processes that drive
normal development and function thoughout the animal
kingdom. Evolution, genetics, and pathways are bed-
rock concepts that inform all our biological investigations,
and they are so much a part of daily life that they form
the subliminal backdrop that the architect must embrace.

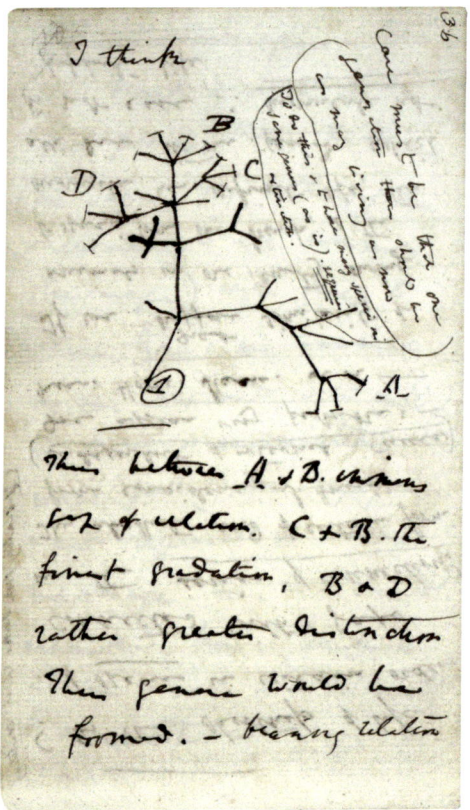

24

Page from Charles Darwin's notebook around July 1837
showing his first sketch of an evolutionary tree

EVOLUTION:
FROM VARIATION TO FITNESS

Charles Darwin departed on the HMS *Beagle* on
December 27, 1831, shortly after graduating with a divinity
degree from the University of Cambridge. He was hired
officially as a naturalist, despite the fact that his expertise
in this area was limited mostly to beetle collecting.
In reality, the irascible and likely depressed captain
of the *Beagle*, Robert Fitz Roy, was looking for a pleasant
companion. By the time of the *Beagle*'s return five
years later from the trip around the world, Darwin's
observations had already begun to shape his theory
of evolution, but he ruminated over how to formulate
these ideas until 1859, when pushed to publish by learning
of a competitor, Alfred Russel Wallace. In *On the
Origin of Species by Means of Natural Selection* Darwin
provided evidence that new species are generated not
by a Creator but rather by selection from existing species.
Even subtle variations in body form and function
over time can affect fitness, when improved, bettering
the chance of survival, and when reduced, leading to
extinction. Of course, the genetic changes that cause the
variations were obscure to Darwin and his contemporaries.

Darwin (1809–1882) at the age of thirty-one

Charles Darwin's ship, the HMS *Beagle* (1912)

Charles Darwin's house in Downe, Kent

Charles Darwin's personal study

Evolution is relevant to architecture in many ways. For both, form fails without function. Decorative male plumage provides an advantage only if the female responds.

Many forms retained through evolution are beautiful in their functional simplicity. The bee's honeycomb wastes no space, generating a structure of hexagonal symmetry that packs the most honey while economizing on the wax used. Sir Christopher Wren designed the Dean's Staircase in Saint Paul's Cathedral based on principles of the spiral of the snail shell.

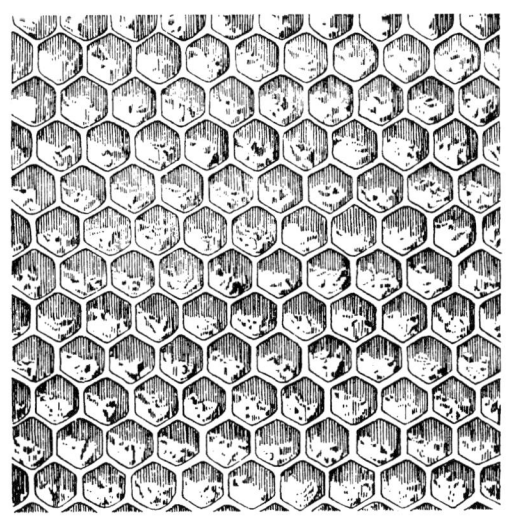

Honeycomb

Dean's Staircase in St. Paul's Cathedral

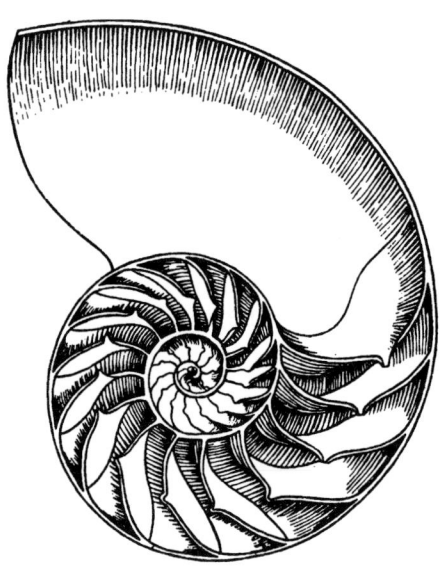

The shell of *Nautilus pompilius*

The lifetime fitness of an animal, or of a building, is
a trade-off between utilization perfected for today and
flexibility in the face of environmental challenge. Built
into living systems at all levels, molecular to organismic,
is a robustness in the face of challenge. It buffers any
change. But if the environment changes, and the form
cannot adapt, the species becomes less fit and ultimately
extinct. Similarly, if labs cannot readily change to new
layouts, they are locked into today's science, and their
inflexibility contributes to programmatic conservatism. 31

Gregor Mendel (1822–1884) in his garden

GENETICS:
CODES AND PATHWAYS

The biological explanation of species variation comes principally from genetics. This field was launched by experiments done by Gregor Mendel in his monastic garden in what is now Brno in the Czech Republic and published in 1866. The report garnered no apparent interest for fifty years. Mendel pollinated pea plants by hand over many generations. He found that the properties of offspring are not a blending of those of the parents. Progeny could resemble one or the other parent, or neither. Mendel proposed, based on statistics of these properties of offspring, that inheritance could be explained by hereditary factors (now called genes), one coming from each parent. The physical basis of the gene was progressively understood over the next one hundred years, as Thomas Hunt Morgan and his colleagues, using the fruit fly, connected inheritance to the nuclear chromosomes (1915); Friedrich Miescher, using human cells and salmon sperm, proved that the nuclear material is DNA (1871); and Oswald Avery, Colin MacLeod, and Maclyn McCarty, showed, using bacteria, that DNA alone, without protein, could change heritable features (1944).

The model for the structure of DNA proposed by James Watson and Francis Crick (1953) led to the understanding of how the sequence of bases in DNA can encode proteins and copy itself exactly during cell division. We learned that the genetic code is universal and that most genes and proteins have close relatives across all animal species, in fact often utilized in similar ways. In other words, nature has proved quite conservative.

Every biological scientist leans on the principles of genetics pretty much every day, both theoretically and practically. We use genetics and the understanding of how genes encode proteins to explain normal functions of cells in the body and how their perturbations make us sick. Much of what is done in laboratories with cells and animals is based on our confidence that the conservatism of nature means that results can be extrapolated, even if cautiously, to humans.

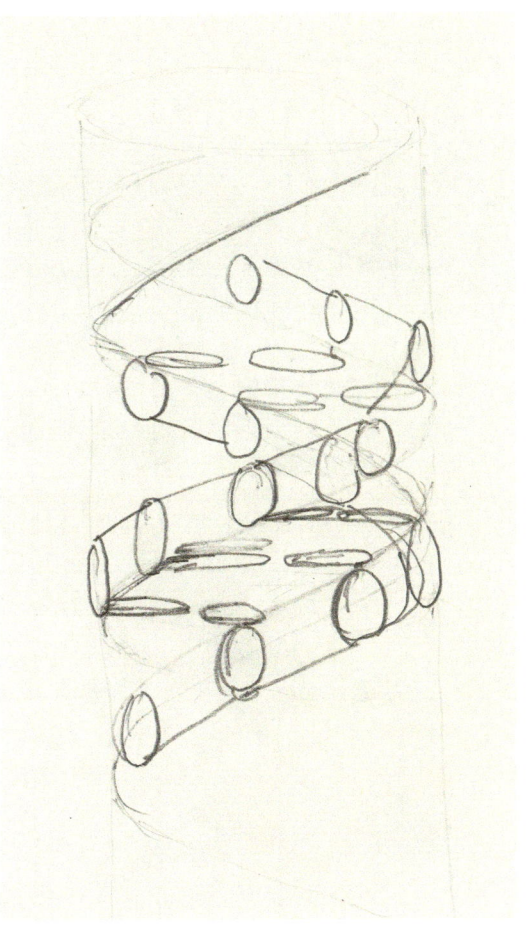

A sketch by Crick, conceptualizing
the structure of DNA as a double helix

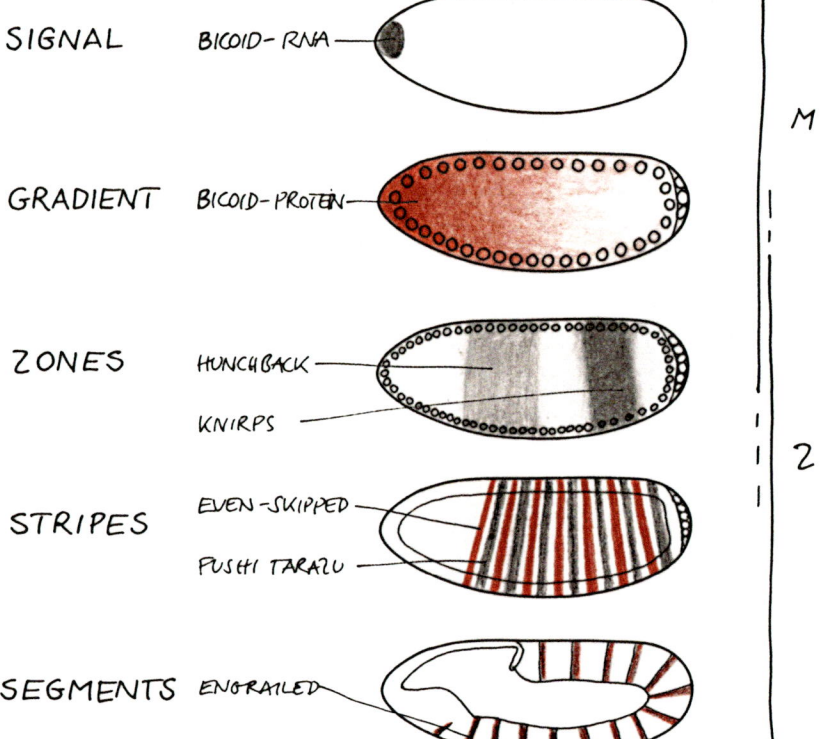

SIGNAL · BICOID-RNA

GRADIENT · BICOID-PROTEIN

ZONES · HUNCHBACK · KNIRPS

STRIPES · EVEN-SKIPPED · FUSHI TARAZU

SEGMENTS · ENGRAILED

M

2

Patterns in the *Drosophila* embryo that led to
understanding of molecular pathways

PATHWAYS AND NETWORKS

Genes encode proteins. Proteins bind to or modify other proteins in stereotyped networks, or pathways. Many protein networks were first elucidated in yeast, worms, and the embryo of the fruit fly, *Drosophila melanogaster*. It turned out that not only were the proteins themselves highly conserved among all of these organisms, but so was their pattern of connectivity. This enabled the under-standing of how protein networks work in humans, how their perturbation can cause disease, and how drugs can be designed to disable, for example, cancer-causing pathways.

THE GARDEN AND THE TAR PIT

THE GARDEN:
EDEN AND FINZI CONTINI

The earliest medicines came from plants. From the beginning it has been known that every medicine has a good and ugly side, a therapeutic and toxic potential. The *Ebers Papyrus*, dated around 1600–1500 BC, with information likely from as early as 3000 BC, contains references to many herbal remedies and notes the potential for toxicity along with palliation.

For more than six thousand years, the milky fluid from the seed head of the opium poppy *(Papaver somniferum)* has been dried into opium to provide pain relief (mostly from the morphine). But if not titrated carefully, opium leads to death from cessation of the drive to breathe, and with regular use causes addiction.

In 1785 William Withering described the beneficial effects of extracts of the foxglove plant *(Digitalis purpurea)* upon patients with heart failure. (Digitalis and its derivatives continued to be used as a mainstay of heart failure therapy until recently.) But Withering also noted that too much digitalis brought on nausea, a slow pulse, yellow vision, and ultimately convulsions and death.

Morphine

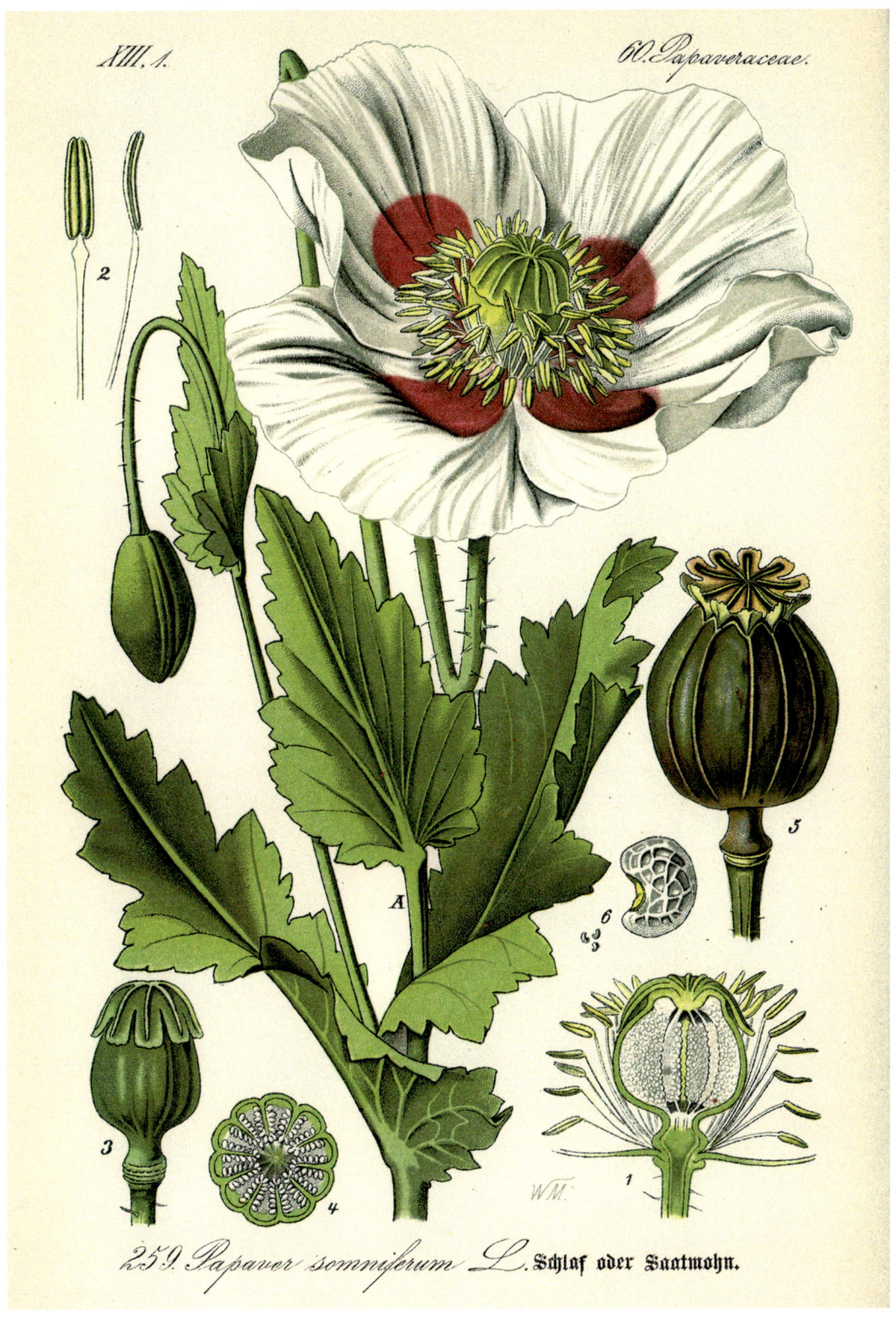

XIII.1.   *60.Papaveraceae.*

51

259 Papaver somniferum L. Schlaf oder Saatmohn.

Papaver somniferum

Digoxin

Tafel 33.

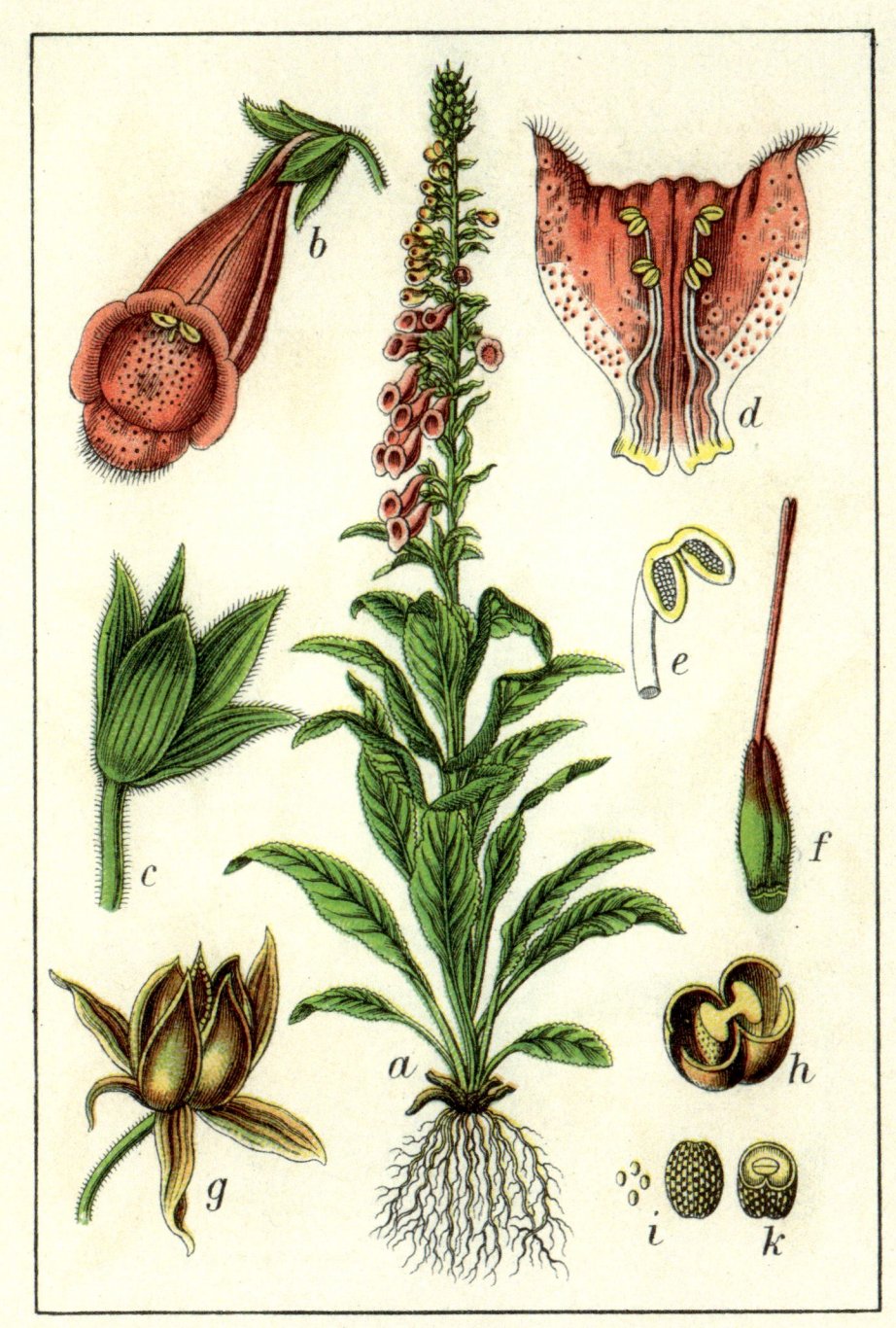

53

Roter Fingerhut, Digitalis purpurea.

Digitalis purpurea

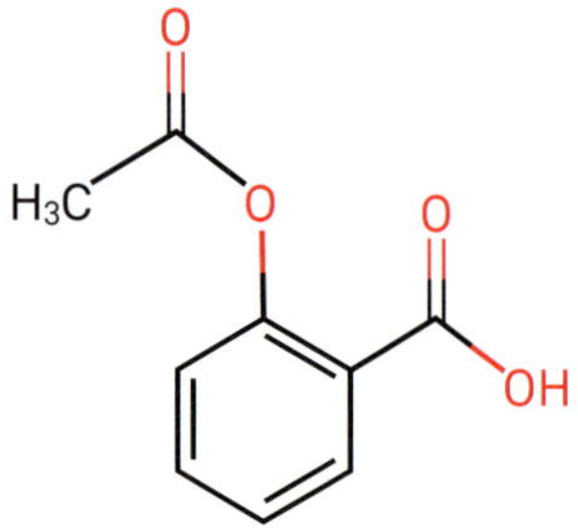

Acetylsalicylic acid

The bark of the white willow tree *(Salix alba)* has been used to treat fever and pains for four thousand years, with the active ingredient isolated in the 1800s and later modified to what is known as aspirin. Today aspirin is used not only as an anti-inflammatory but also for the prevention of cardiovascular disease. However, its overdose leads to gastric ulcers and bleeding.

Tafel 23.

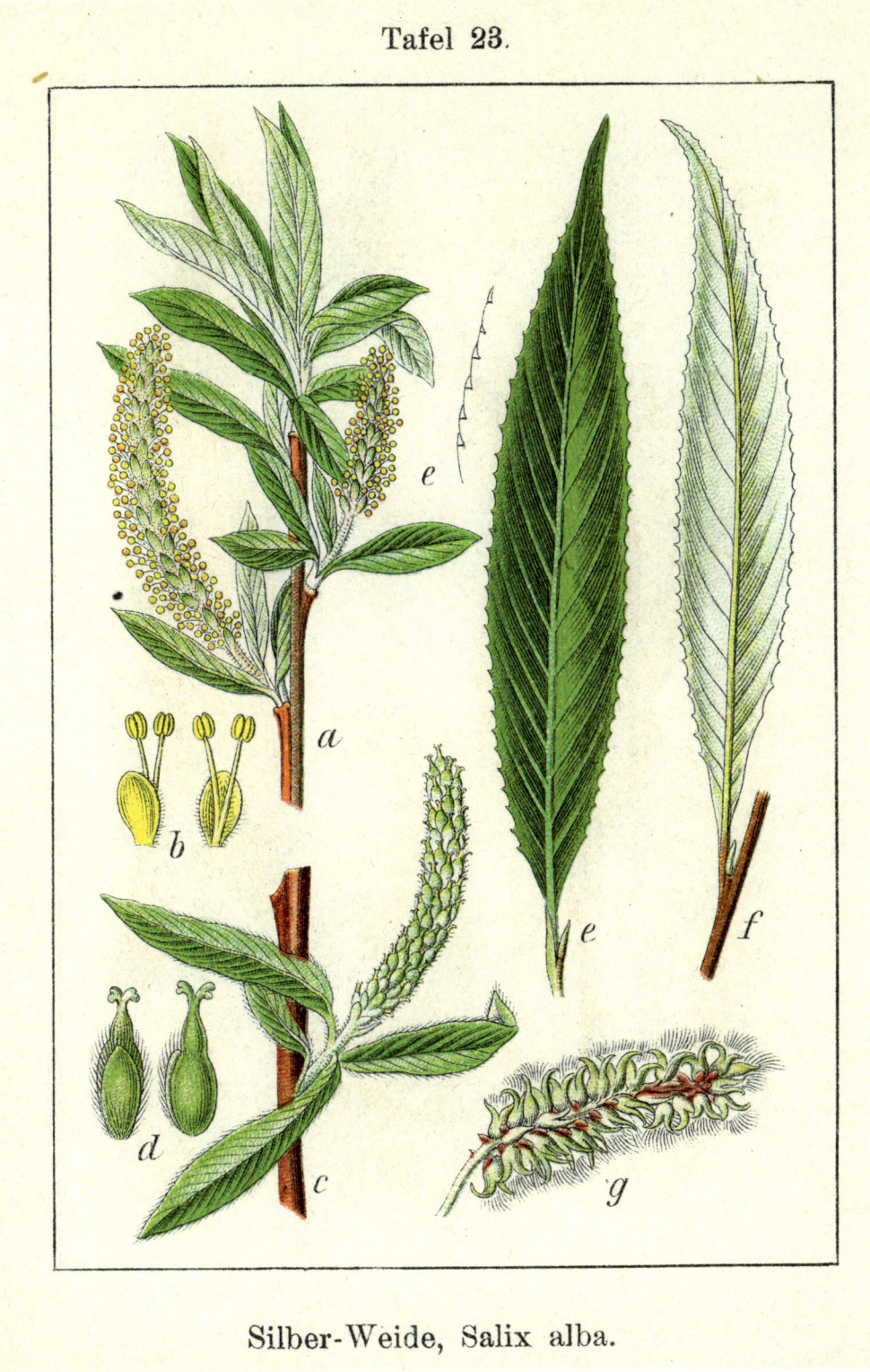

55

Silber-Weide, Salix alba.

Hence, the strange dichotomy of the garden: it provides sustenance and health along with threats. All drugs, without exception, embody this knife-edge.

Gardens are part of both campuses, to remind us both of pharmaceutical origins and the balance between the life-giving and toxic effects of medicines. As important, the gardens serve as sites of escape and contemplation. In Shanghai especially, we have borrowed elements of the traditional Chinese garden. Winding paths lead to surprise and revelation as a corner is turned; varieties of plants provide a palette of color and texture that varies with the season; meticulously positioned rocks and stones provide the essence of great mountains and rolling hills even though the garden is not at the scale of the original imperial landscapes.

Chinese garden

In science we are working at the edge of what is known or believed possible. So how, architecturally, should we define limits? Stone walls are a distinctive element of the New England countryside. These walls had their origins as piles of rocks dragged to the edges of property by farmers clearing their land. Later they were built intentionally, but still without mortar, and so are discontinuous along their length and imperfect by design. They cannot and do not exclude movement and exploration. As Robert Frost said in "Mending Wall," "Something there is that doesn't love a wall / That wants it down." It is this peekaboo element that makes them inviting rather than a barrier.

New England stone wall

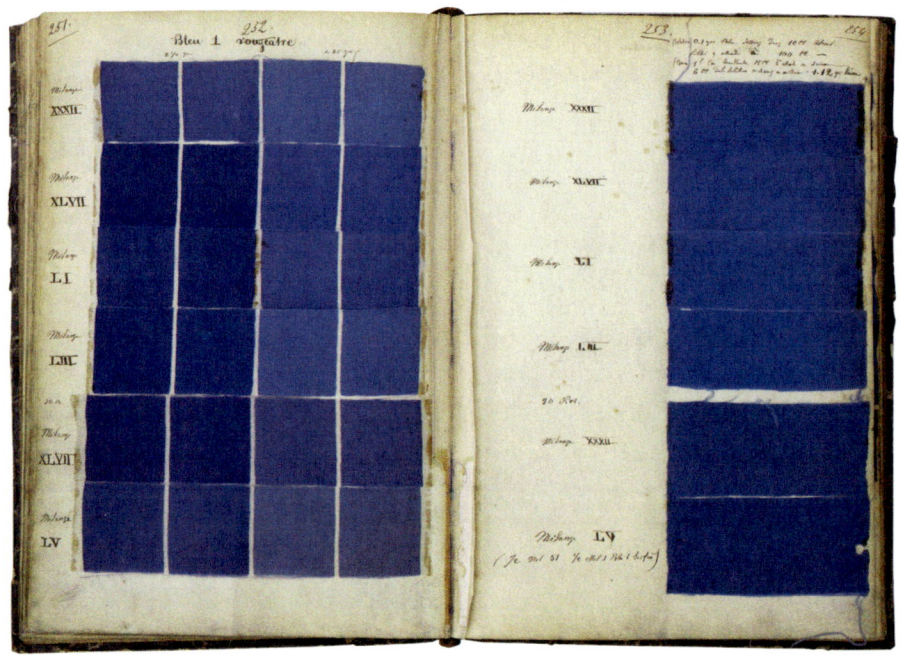

Prints of aniline blue from a manufacturing
inspection register, Switzerland, 1862

COAL TAR:
FROM DYES TO MEDICINES

The gas for the streetlights of Victorian England
derived from the burning of coal, which left a disgusting
black gooey residue called coal tar. In 1856 an eighteen-
year-old chemist, William Henry Perkin, discovered to his
surprise that he could extract from coal tar a colorfast
purple dye, termed mauve, or aniline purple, which could
stain silk and other fabrics. Coal tar, it turned out,
was a collection of many molecules, including many new
dyes. Until then dyes had been extracted only from plants,
a laborious and expensive procedure. This discovery
launched an international race for new dyes and led to
the formation of new chemical companies.

About fifty years later, another major discovery led to the transformation of many of these chemical companies into pharmaceutical companies. On August 31, 1909, Sahachiro Hata, a bacteriologist visiting Paul Ehrlich, the director of the Royal Institute for Experimental Therapy in Frankfurt, found that compound no. 606 cured a rabbit infected with syphilis. The sores regressed and the responsible spirochetes disappeared from the blood. By the next year, compound no. 606 had been tested and shown to be effective in clinical trials, and was marketed as Salvarsan by the chemical company Hoechst. It remained the cure for syphilis until antibiotics were discovered many years later.

Paul Ehrlich (1854–1915) with Japanese
bacteriologist Sahachiro Hata (1873–1938),
about 1910

Ehrlich had reasoned that he could find a chemical, a "magic bullet," to kill microbes but not the patient, because he and others had discovered chemical dyes that specifically labeled bacteria. He speculated that this staining was due to selective binding of the dye to receptors present on microbes but not present in human cells. He then screened arsenic-bound dye molecules on the assumption that the dye could bring the toxin to the bacteria. Compound no. 606 was the 606th such molecule examined, one by one, in a model of syphilis in rabbits.

Salvarsan thus was the first drug discovered utilizing an underlying biological hypothesis tied to chemical screening, which remains to this day the way new drugs are discovered. It was also one of the first drugs made by chemical synthesis rather than by isolation from a plant. (The use of chloroform as an anesthetic for humans had been pioneered by the Scottish physician James Young Simpson. Chloral hydrate was synthesized as a sedative. Acetylsalicylic acid was generated by Felix Hoffmann as a chemical improvement upon the natural product (salicin) obtained from willow bark and was marketed as Aspirin by Bayer starting in February 1899.)

Hata and Ehrlich's discovery suggested a rational approach to drug discovery, leading Hoechst, Ciba, Geigy, Sandoz, and Bayer, among others, to begin the transition from chemical to pharmaceutical companies.

Of course, the technologies used today have changed.
In many cases screening has been replaced or
complemented by computer-aided drug design based
on the atomic structure of the target and potential
drug, using tools such as X-ray crystallography and
nuclear magnetic resonance spectroscopy. These allow
drug candidates to be designed and changed, atom
by atom, in a rational fashion. So while chemists still
synthesize new molecules in protective fume hoods,
much more of their time is spent in front a computer.
Chemical reactions are often done in smaller quantities
because analyses require miniscule amounts of
reagent. And robots can do much of the repetitive work
of testing various chemicals. In many cases today
the drug is not a synthetic chemical but a biological
reagent—a protein, for example. But Ehrlich's
combination of biological insight coupled to screening
and chemistry remains the foundational approach to
drug discovery.

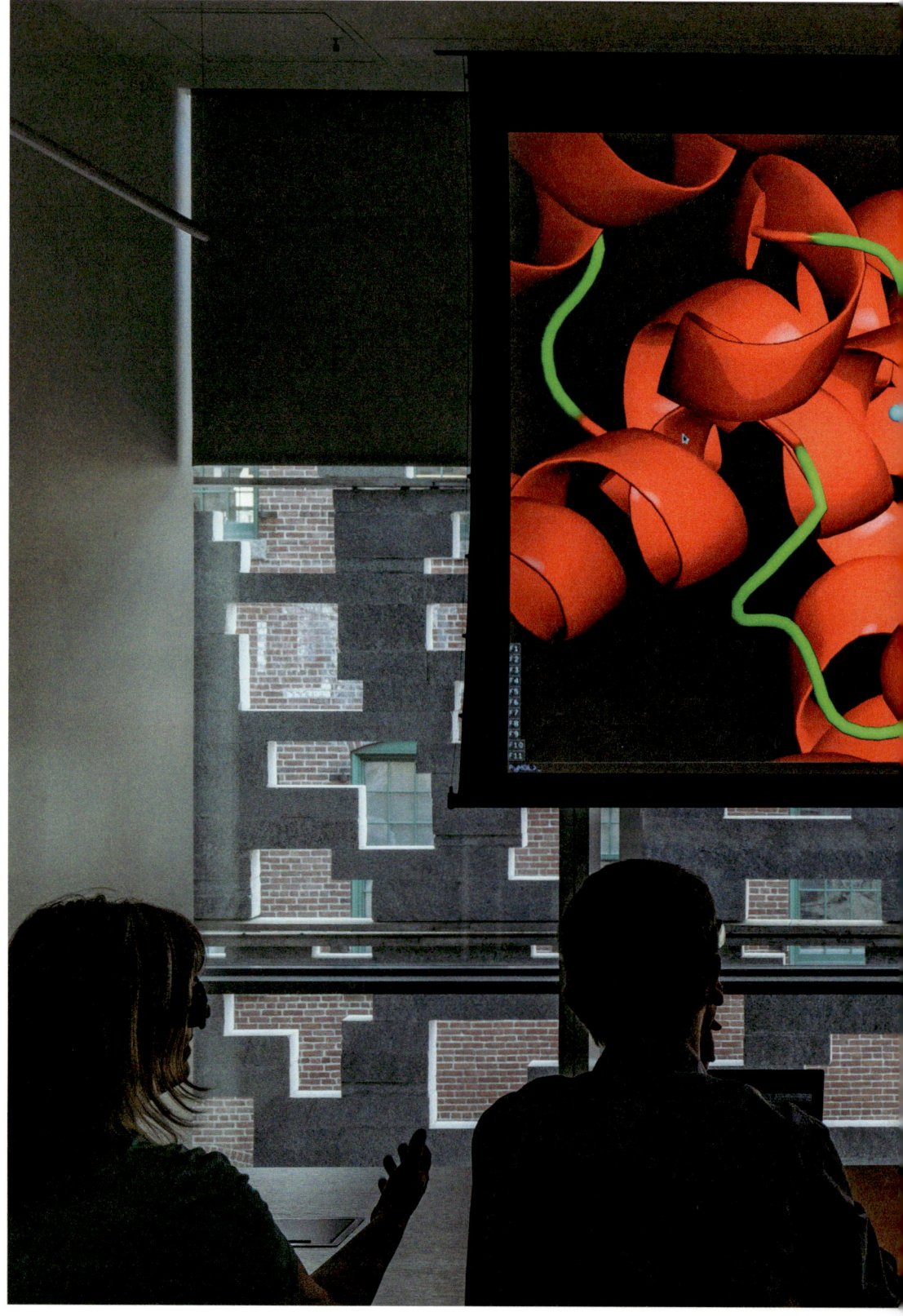

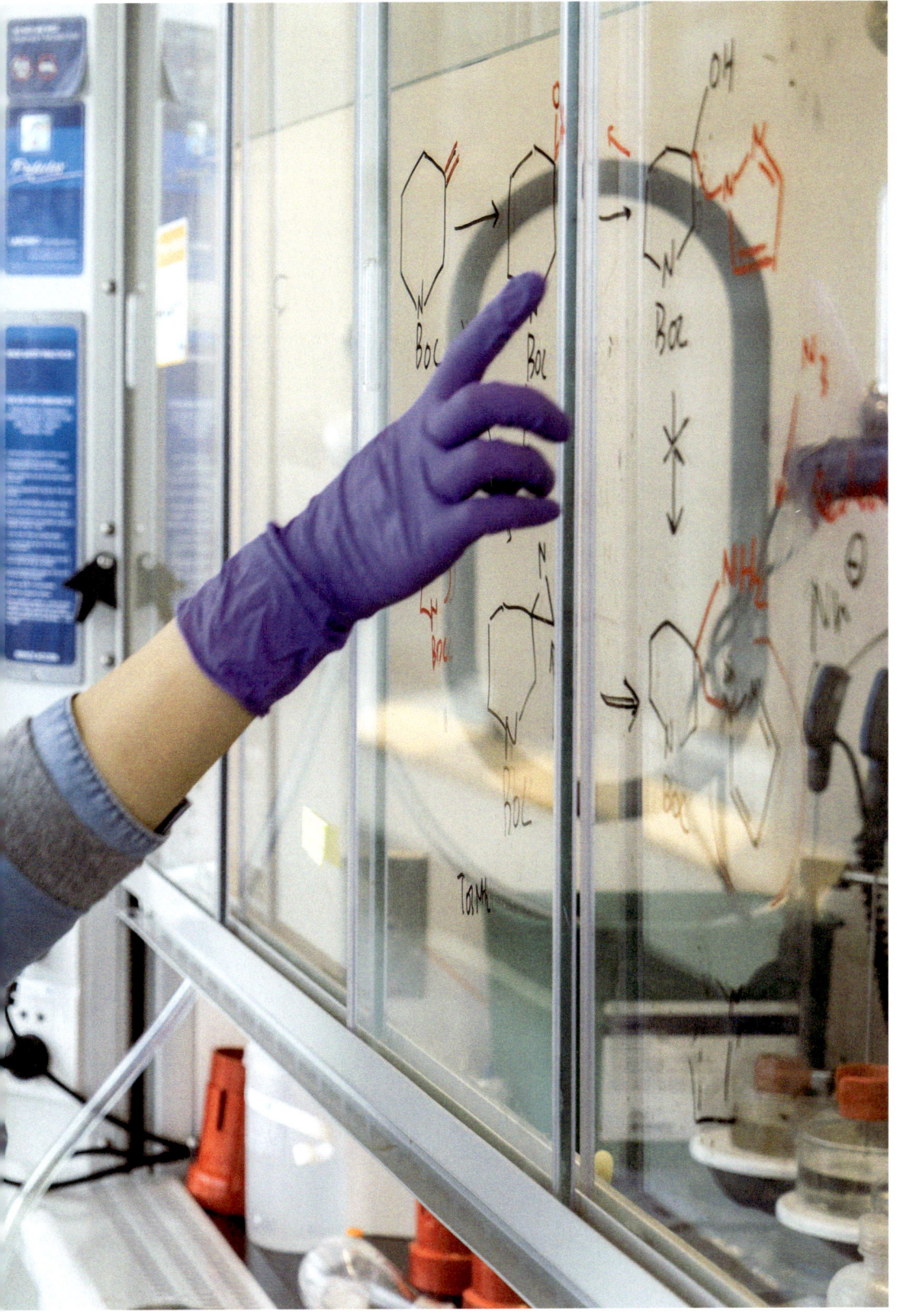

THE UR LAB
FROM ALCHEMY TO SCIENCE

Al-Razi with lab equipment

ALCHEMY: LABS BEFORE SCIENCE

Dedicated scientific workshops appeared with the alchemists who used chemical and biological experimentation as they sought the philosopher's stone. Some of the first alchemists were Islamic scholars, such as Jabir ibn Hayyan (ca. 722–815), who incorporated elements of mysticism, religion, astronomy, and mathematics into their alchemical attempts to generate gold from base metals. Abu Bakr Muhammed ibn Zakariya al-Razi (ca. 865–925) was more practical, describing in detail early laboratory equipment and scientific processes such as distillation.

Alchemy was introduced to Europe via translation of the Arabic *Book of the Composition of Alchemy* in 1144. Like their Arab predecessors, European alchemists based their work on Aristotle's theory of the four elements of matter (earth, air, fire, and water), and their goal was generally to transmute base metals into gold.

We have only a poor idea of what these labs looked like, because alchemists worked in secrecy, and their approaches were as much mystical and theological as technical. The images of alchemists as individuals lost in thought near their furnaces, sometimes with a few assistants running the bellows, probably are fanciful, since as far as we know, alchemists did not invite outsiders to their lairs and shared little of their information. It is likely that their labs generally centered around a furnace and utilized a distillation apparatus to separate

Roger Bacon in his lab

84

Paracelsus in his lab

chemicals. These in turn mandated some attention to safety and ventilation.

Their motivations varied. Many were charlatans, using vague mystical explanations and magical incantations to extract funding from wealthy patrons. But others, such as Roger Bacon (ca. 1219–1294), a Franciscan friar, turned to lab experimentation to understand and confirm the truths contained in the Bible.

The Renaissance alchemist, scholar, and physician known as Paracelsus (1493–1541) believed that experimental chemistry tools of alchemy should be turned not to making gold but to the service of medicine. This use of chemistry to refine potential medicines distinguished him from the apothecaries of the day, who used mixtures of herbal remedies. Some point to Paracelsus as the first medicinal chemist. He realized the importance of careful dose titration and thereby is considered to have begun the field of toxicology.

Robert Boyle doing a demonstration of an experiment

EXPERIMENTAL SCIENCE: OPENNESS AND REPRODUCIBILITY

More careful experimentation as the sole guide to scientific truth, without accompanying theology, philosophy, and mysticism, became a standard approach only in the late sixteenth century. William Harvey (1578–1657) demonstrated the circulation of the blood using human anatomical observation and a cow's heart. Robert Boyle (1627–1691) was one of the first experimental chemists; he discovered the relationship between the pressure and volume of gases, now known as Boyle's Law. Although he too sought in his alchemical labs methods to generate gold, he had moved away from Aristotelian acceptance of the four elements and relied more upon experimentation and analysis. He is viewed by many as being the father of modern chemistry, beginning with his book *The Sceptical Chymist* (1661). Experimentation, according to Boyle, should be reproducible by others, an approach very different from that of the secretive alchemists.

Libavius' (1555–1616) ideal laboratory from the rear, 1606

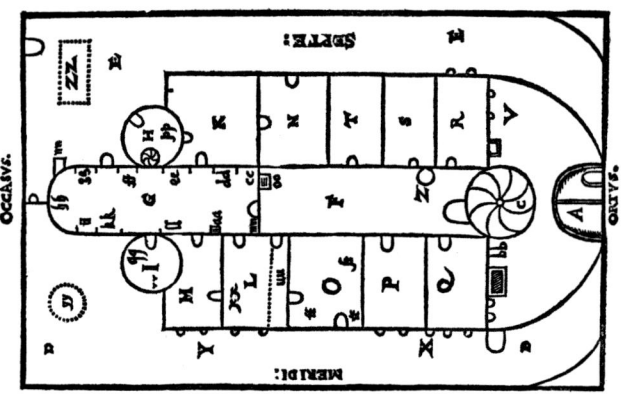

Plan of Libavius' ideal laboratory, 1606

BUILDINGS DESIGNED
FOR SCIENCE

The first extant design for a building dedicated to chemical sciences is from 1597, by Andreas Libavius. It was intended as the home for the chemist and his assistants, and it included rooms for specialized procedures (such as weighing chemicals and for crystallization) and a small private laboratory in addition to the main one, with furnaces distributed for particular activities.

Lavoisier's (1743–1794) laboratory in the Paris Arsenal, ca. 1790

Over the ensuing years there continued to be a move
away from the secretive labs and operations of the alchemists to an emphasis upon rationality, reproducibility,
and openness. Antoine Lavoisier's books, such as *Elements
of Chemistry* (1789), in particular marked this change.
The illustrations of Lavoisier's experiments on oxygen and
the nature of respiration show a lab organized for the
specific experiment at hand: shelving for chemical and
apparatus storage, and tables designed for experimentation. Unlike the private solo operations of the alchemists,
the lab is filled with observers and recorders.

Exterior of the new chemistry laboratory building at Heidelberg
with castle in the background, 1858

By the mid-1800s, especially in the German states, new buildings were designed specifically for chemistry experimentation. They were important in bidding wars between cities to garner the most famous chemists. The chemical institute in Heidelberg built for Robert Bunsen (the inventor of the Bunsen burner), for example, included central provision of services such as gas, electricity, and water, and separate rooms for specialized procedures. The furnace at the center of the alchemist's lab was replaced by central steam, both for heating the lab and for experimentation.

Bench and bottle racks at Leipzig University, 1868

The canonical lab layout standard today is evident already in the Berlin laboratory building constructed in the 1860s for Wilhelm Hofmann. Two-sided symmetrical benches are arrayed in parallel along a central aisle. Each bench has shelving down the middle, separating the two symmetrical halves, and cupboards under-neath. Scientists at the benches work back-to-back. Smelly or dangerous experiments are done in fume hoods provided with updraft ventilation and a glass wall in front to protect the chemist. Separate rooms hold delicate or sensitive instrumentation.

Louis Kahn's (1901–1974) drawing of the convent
of Saint Francis of Assisi, Italy, 1929

Jonas Salk (1914–1995) and Louis Kahn at the Salk Institute

There have been recent attempts to address more than the strictly functional aspects of lab design. The Salk Institute, designed by Louis Kahn in San Diego, is most elegant in its attention to the need for contemplation in science. The monastery of Assisi served as an inspiration for this magnificent building floating above the Pacific Ocean on the Torrey Pines Mesa.

This was one of the first attempts by a modern architect to improve upon the essential lab design basically unchanged since those of the mid-1800s. (Kahn had previously designed the Richards Medical Research Laboratories at the University of Pennsylvania, but this was less distinctive and, most felt, not tremendously functional.) At the Salk, each fellow had a private study for reflection, warmly decorated with oak floors, tables, and bookshelves. The original plan for a central garden was replaced by a plaza of travertine limestone with a water feature, surrounded by a cloister redolent of medieval monasteries.

97

In our case, we felt that the dedication of private studios for all senior staff would move away from the essential openness we desired, and that all scientists, even at the junior level, prefer a variety of opportunities for spaces of quiet and escape. Hence we have offered many types of private areas: some rooms, some simply isolated corners, and some specifically designed nooks near the labs, as well as gardens, rooftops, and patios. The goal is no different from Kahn's—to provide for all needed areas of quiet and contemplation—but to do so in many distinctive settings.

TRAJECTORY:
BUILDING FOR EVOLUTION

Of course it is infeasible to predict where labs will
be in twenty years. But we can make some extrapolations.
One is that the walls and furniture will become akin
to smartphones, part of the communication systems,
effectively rendering much of the space a drawing board
for planning and data analysis. I have been in old
labs where the walls are covered with the scribbling that
accompanies experiments, and today the glass fronts
of chemical hoods generally are replete with structures
and plans for experiments. Why not have these be
available to colleagues across the ocean and linked
to the lab notebook?

It also seems likely that the open lab bench will become
a smaller part of the plan as more and more exploration
is done using very specialized devices and computational
analysis. There likely will be even more blurring of lines
between today what is deemed chemistry and biology,
and as chemical experiments are miniaturized, protective
chemical hoods will be replaced by tiny modular devices
taken anywhere for temporary use.

It seems likely that the tremendous variety of person-alities engaged in the scientific enterprise will be quite recognizable even centuries from now. There will be roles for both the quirky individualist and the technocrat, for the soloist and team player, for the flexible and the rigid, and for the introvert and the extrovert. Motivations of scientists are as variable as in any profession. But their work renders the world ever more comprehensible and livable, so we should make their homes ever more beautiful, functional, and inspirational.

CAN BUILDINGS DRIVE SOCIOLOGY?

ROWING SOLO OR ON A TEAM

Isaac Newton and Robert Boyle believed they were members of a select chosen few favored by God for revelation of His mysteries. But all evidence suggests that it is rare, indeed, for a great discovery to be from a single scientist, working in isolation, without refinement from a collaborator or competitor. Even Einstein needed mathematical advice from his friend Marcel Grossmann on the way to the general theory of relativity. The tension between two minds wrestling with the same data often provides the route to clarification. Furthermore, in many fields large teams drive investigations, from human genetics, which requires expensive technologies and large populations from all over the world, to the discovery of new medicines, which crosses many disciplines of biology, chemistry, and medicine.

But once groups are involved, social behaviors set in. Species evolved social behaviors for mating, feeding, and protection from predators. With collective behavior came, for some species, including humans, competition. Issues with leadership and aggression can determine success and failure in science. Communication between groups of different instincts and backgrounds can lead to the greatest discoveries but can also cause friction.

So how do we accommodate individual instincts, ambitions, and drives along with teamwork? Can space somehow be used to open communication and flatten hierarchies? Can it be comfortable both for the ambitious individual and for collective activities?

105

OPENNESS AND CONNECTIVITY

In both Cambridge and Shanghai, space is now open, without dedicated offices or closed labs. There is no longer the opaque door as an energy barrier to interaction. We eliminated the visual segregation across benches brought about by the glassware-containing central shelves that have been standard since the origin of classical German laboratories, and we opened the views to take in broad window expanses. (Storage now is centralized on each floor and maintained by bar code, so there is no need to sequester it in individual areas.) Corridors are broad and inclusive of break areas for congregation and coffee.

It is known that interaction falls off rapidly after ten meters of separation, and even more so between floors. So we have tried to build in stairs as integral features of the buildings. In the Toshiko Mori building, for example, the stairwell forms the central spine of the building, its very openness to the outside and within cementing interaction between floors.

In the Maya Lin building, coffee stands and broad stairs near the entrance become meeting points for casual conversation, reminiscent of the Spanish Steps in Rome.

The terraced bamboo stairwell of Yung Ho Chang embraces informal open meeting platforms.

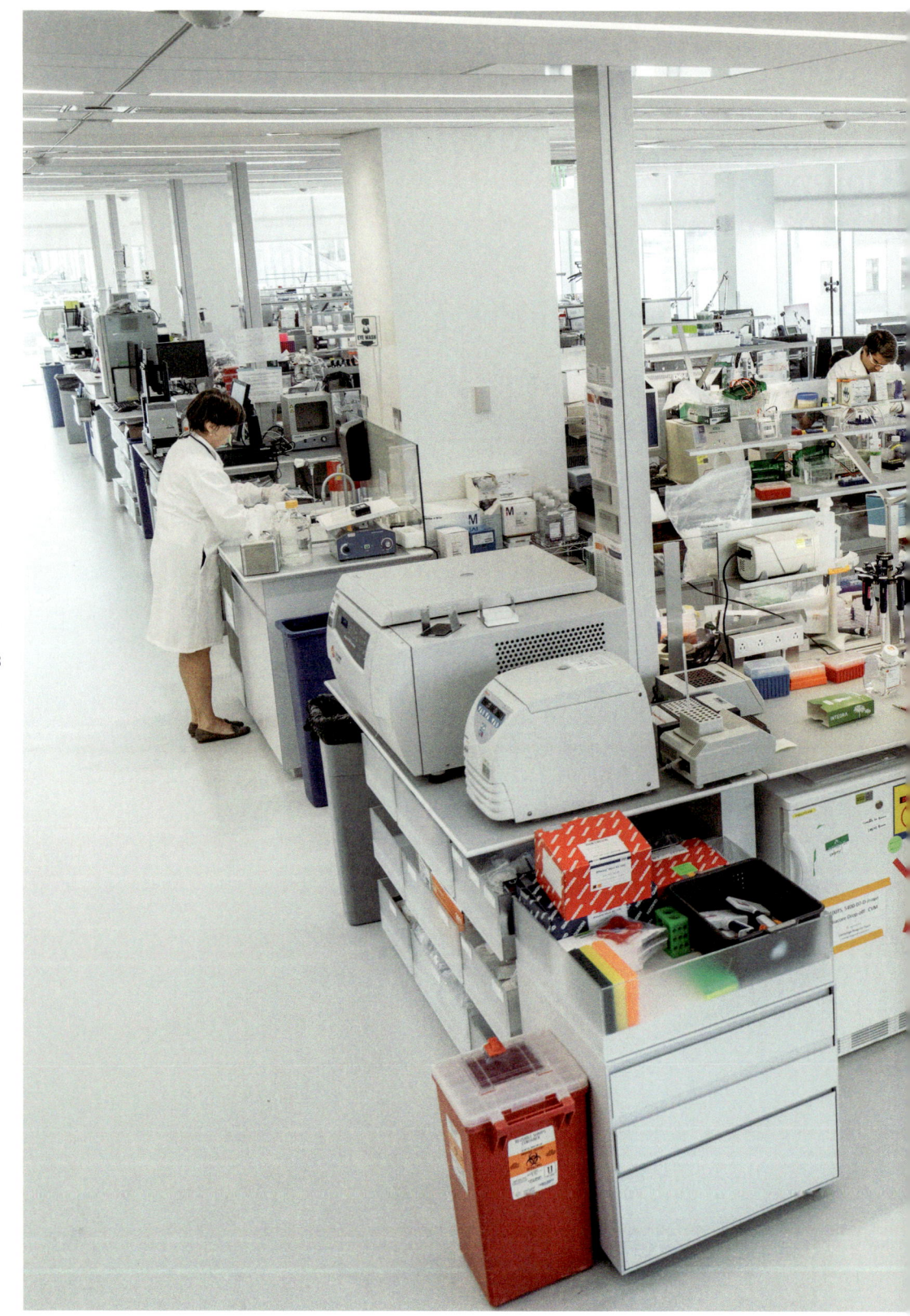

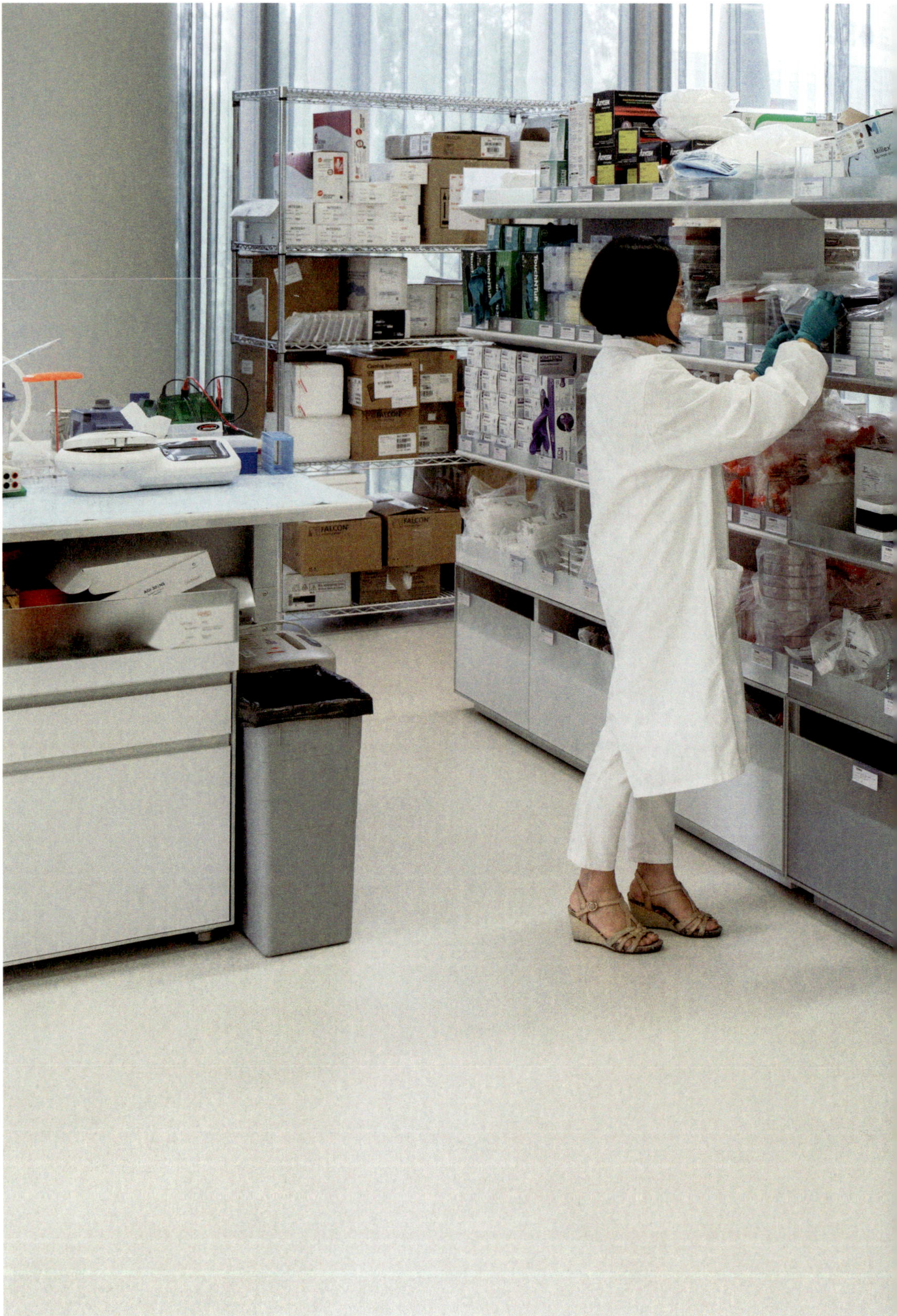

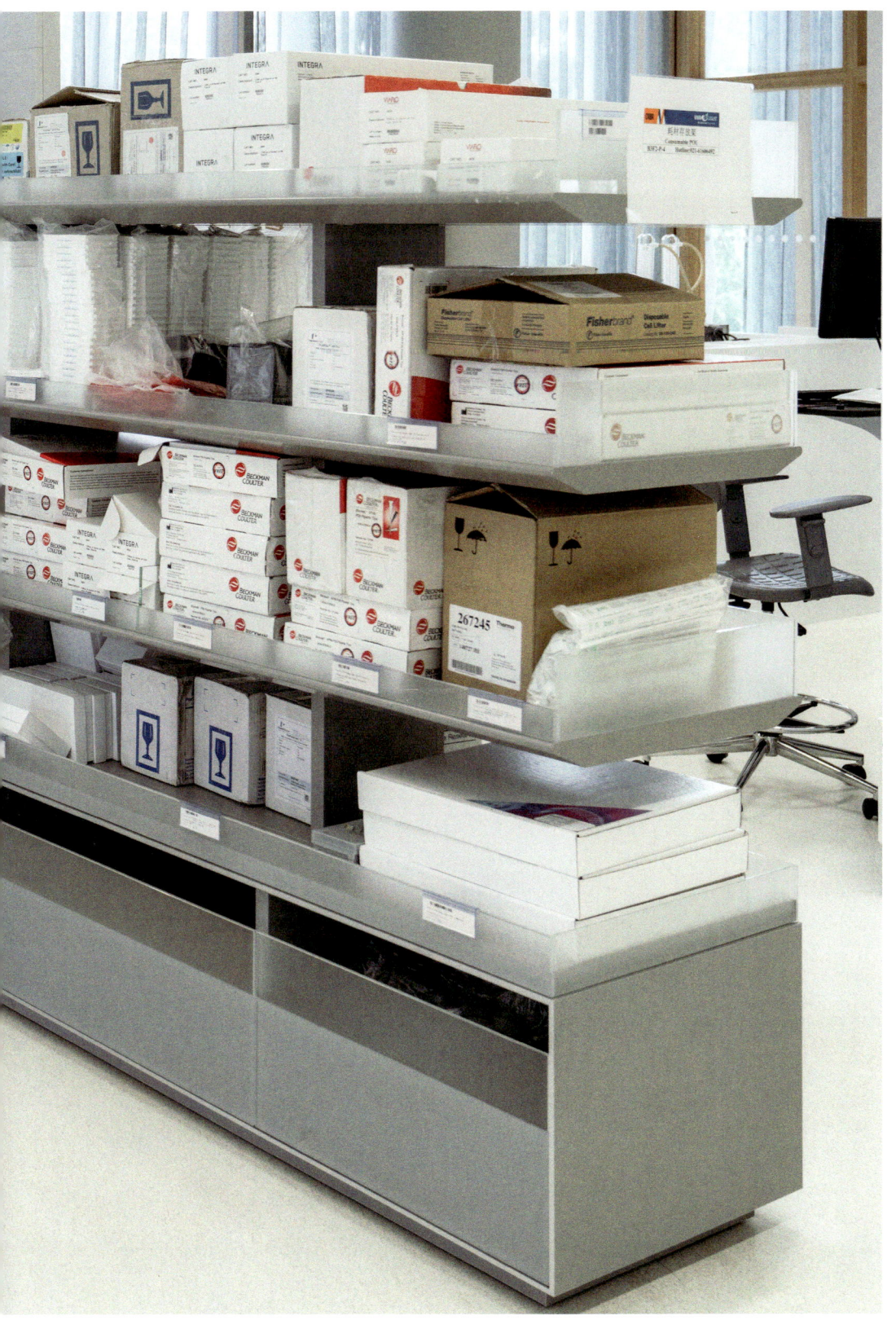

GIMME SHELTER

Many scientists are introverts, unable to concentrate
if forced to listen to music or conversations of neighbors.
Most of us range between intro- and extroversion
throughout the day, depending on internal clocks and
emotional states, and everyone wants quiet space
for contemplation on occasion. Thus, much of the campus
is dedicated to the provision of space for escape, either
solo or in small groups. These include quiet rooms, small
nooks around corners, gazebos, benches, and tables
on the lawn, booths open or with the occupant hidden,
and simple high-backed chairs facing away from the
action out onto an atrium.

Pausing for coffee or a snack is key to scientific
communication and relaxation. Coffee areas are magnets
for the tired, the drained, the overexcited, or the
lonely. Many spots are designed to be in easy reach
of coffee or snacks. Cafeterias are central meeting
points designed both for private lunches and for groups
of all sizes.

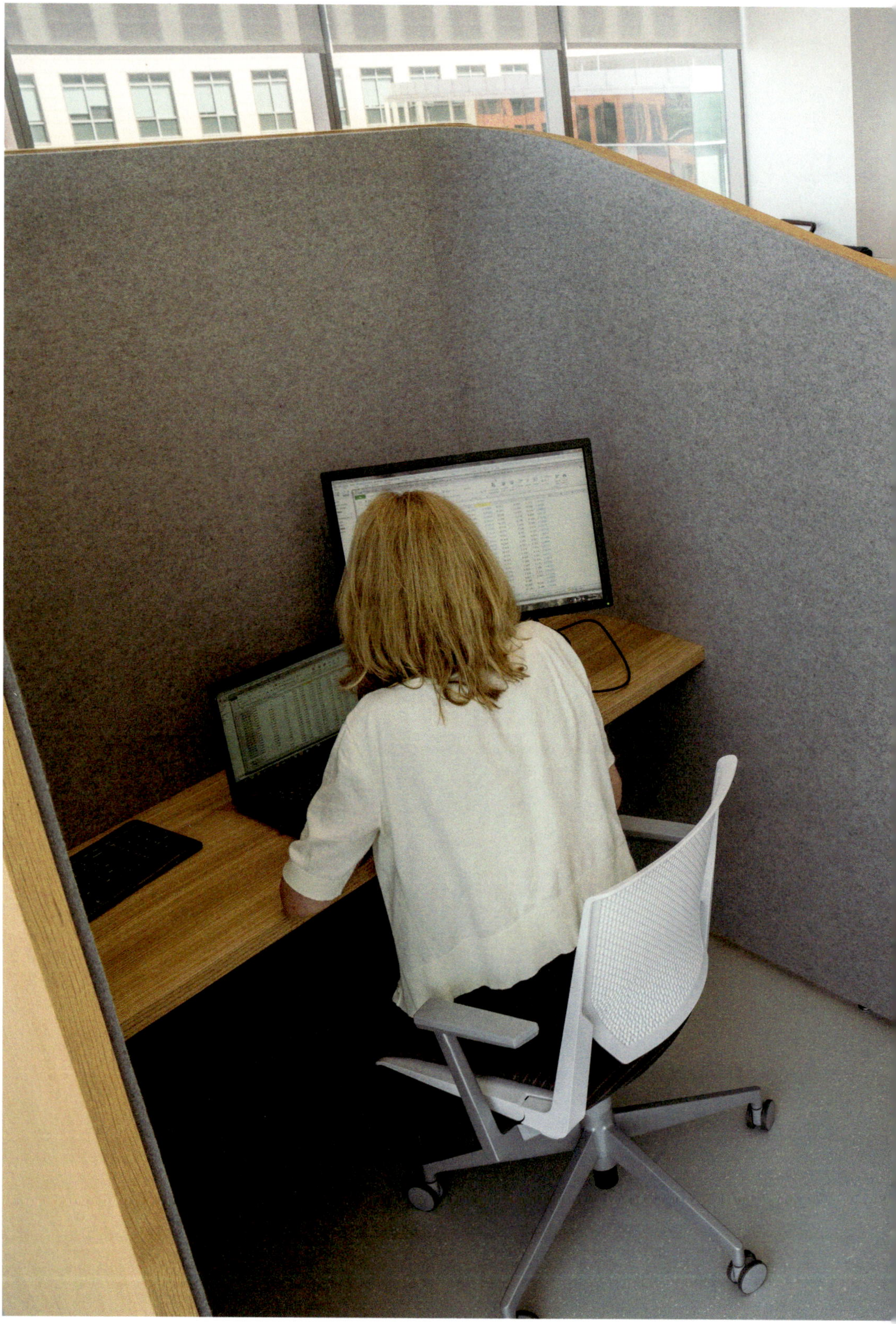

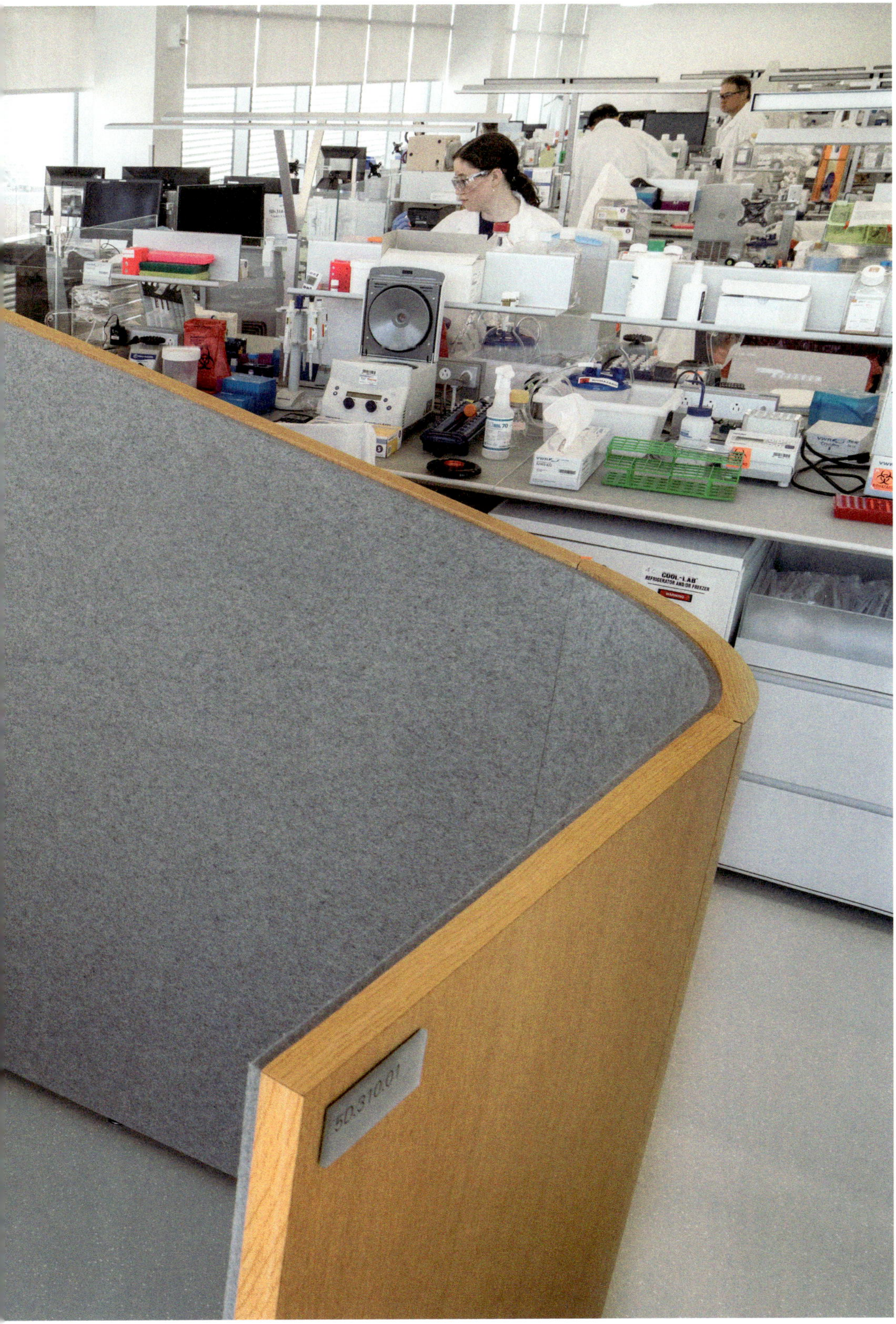
COOL-LAB
REFRIGERATOR AND/OR FREEZER
5D.310.01

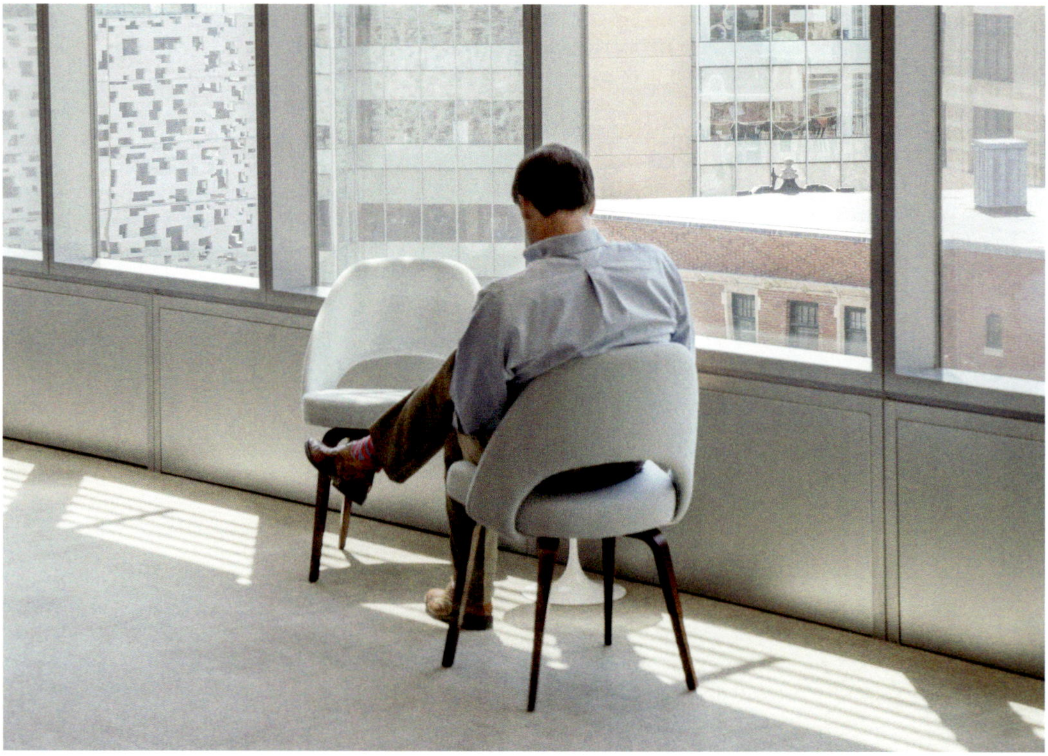

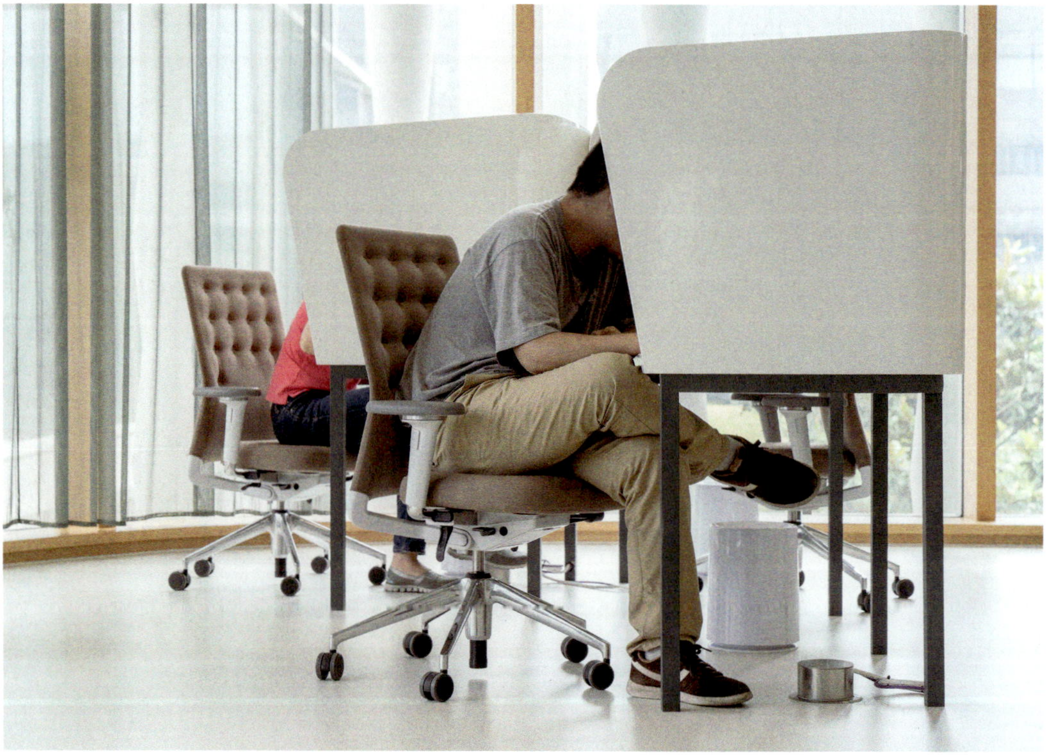

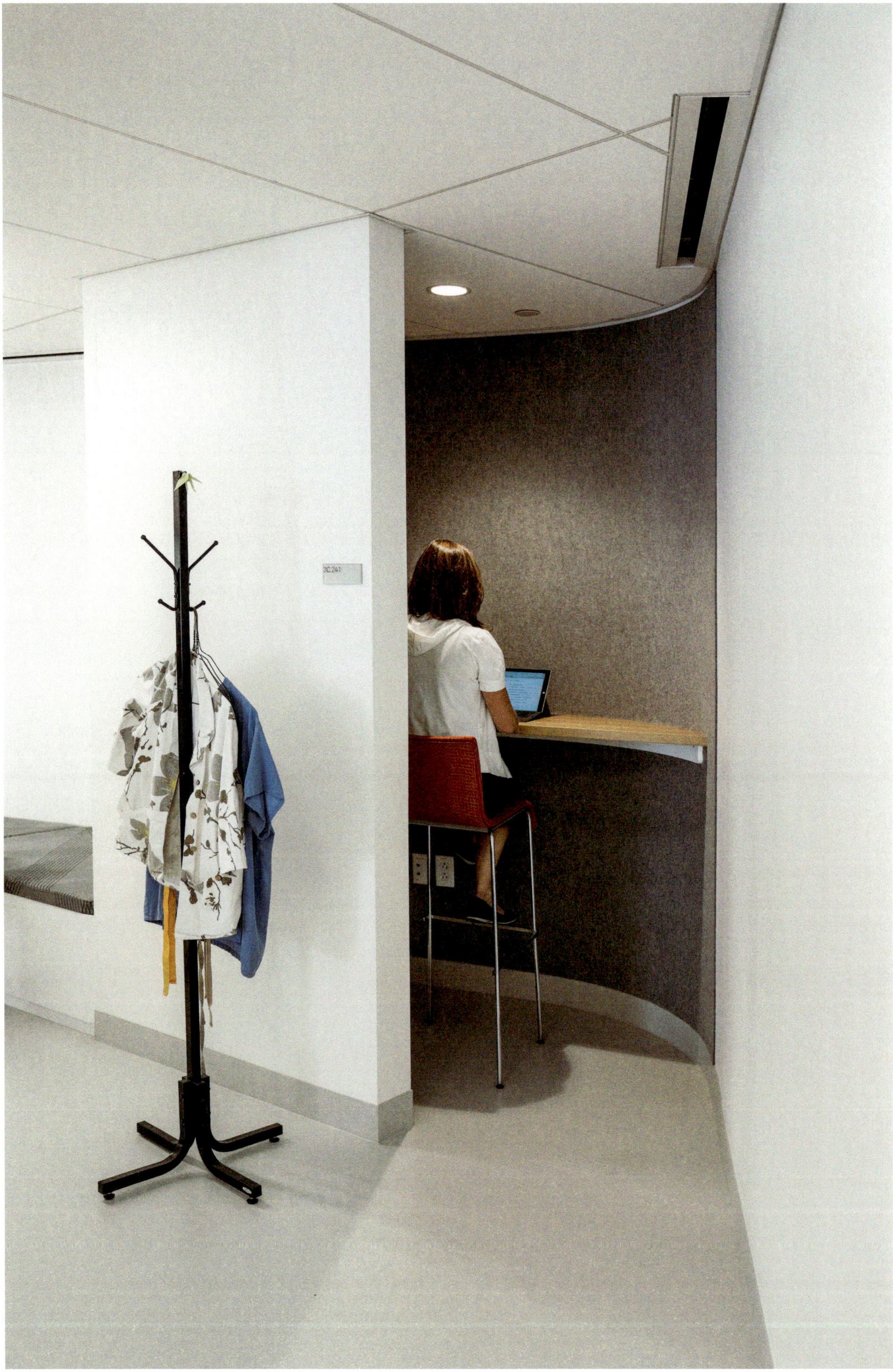
3C.241

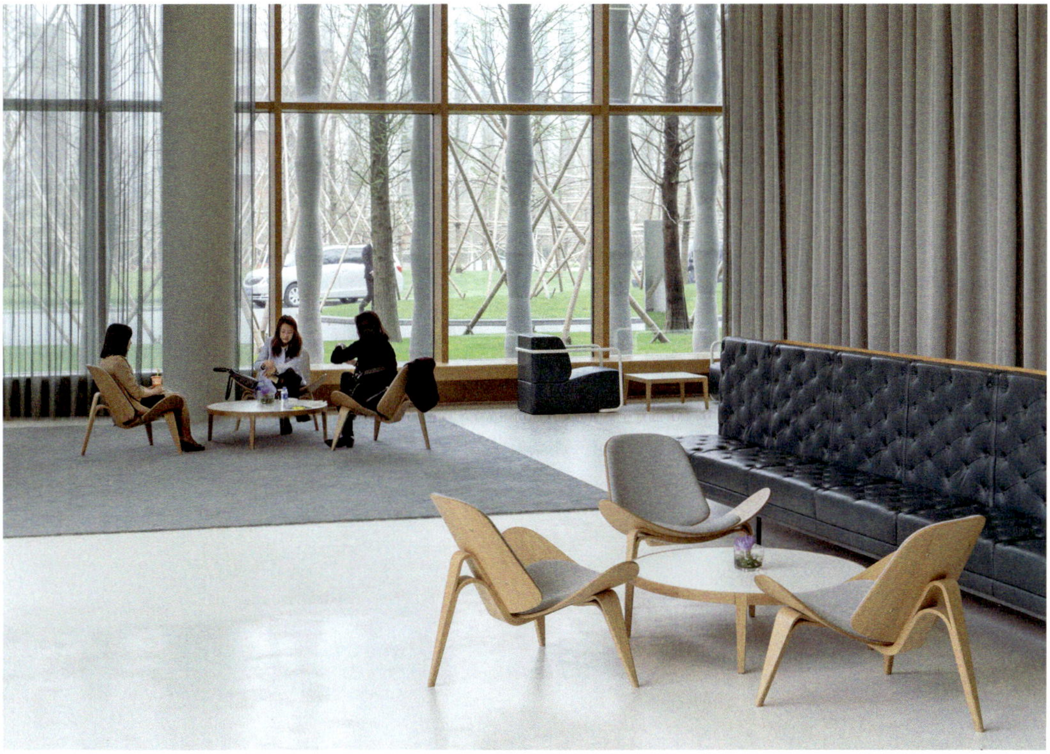

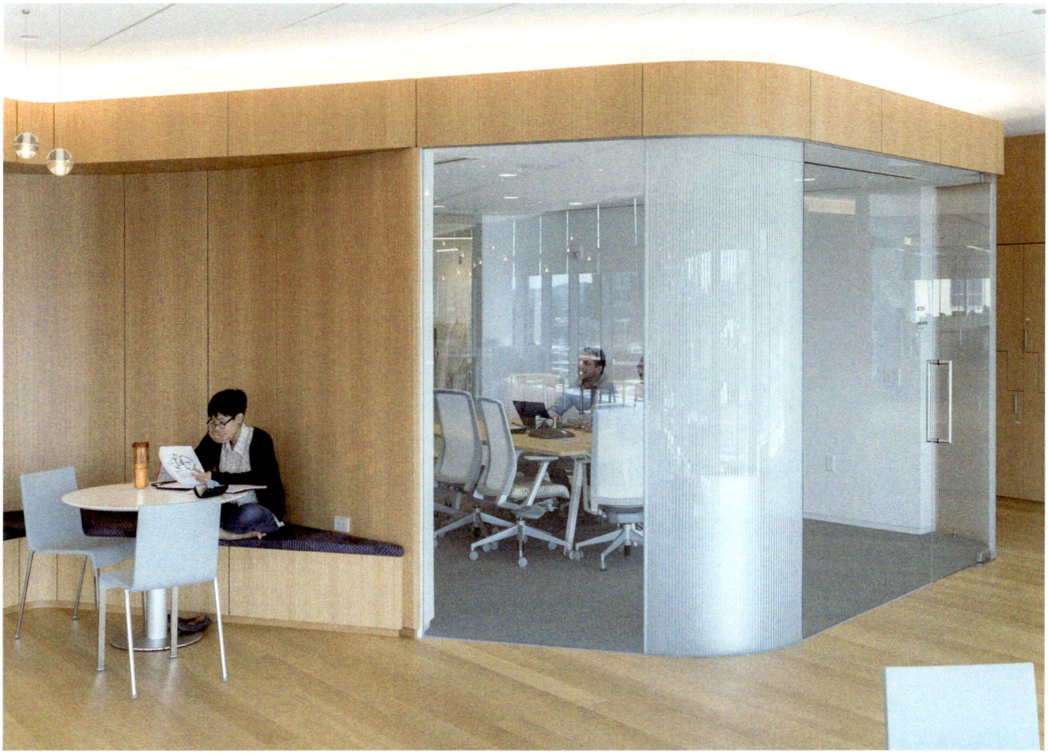

MODULARITY:
THE "LEGO" LABORATORY

Fortunately we no longer need the alchemist's furnace.
But the standard lab bench constrains changes in
program and renders even minor renovations inordinately
expensive. A lab bench is not only built in but also
is attached permanently to plumbing and electricity.
We wished to design a lab that can evolve, that can
be readily redesigned by the scientists themselves as
needed. It must accommodate needs for heavy equipment
and the storage of plasticware and reagents. And of
course it needs to be elegant and incorporated into an
open-space plan.

With Toshiko Mori we designed a system of furniture
that is completely modular, click-and-play. Structurally,
all utilities (telecommunications, vacuum, electrical,
etc.) are provided though the ceiling, with vertical chassis
at regular intervals. These can be connected to "spines,"
the backbone of lab benches, desks, or structures to hold
equipment.

Little storage is above eye level, making the lab open to
sunlight and communication throughout.

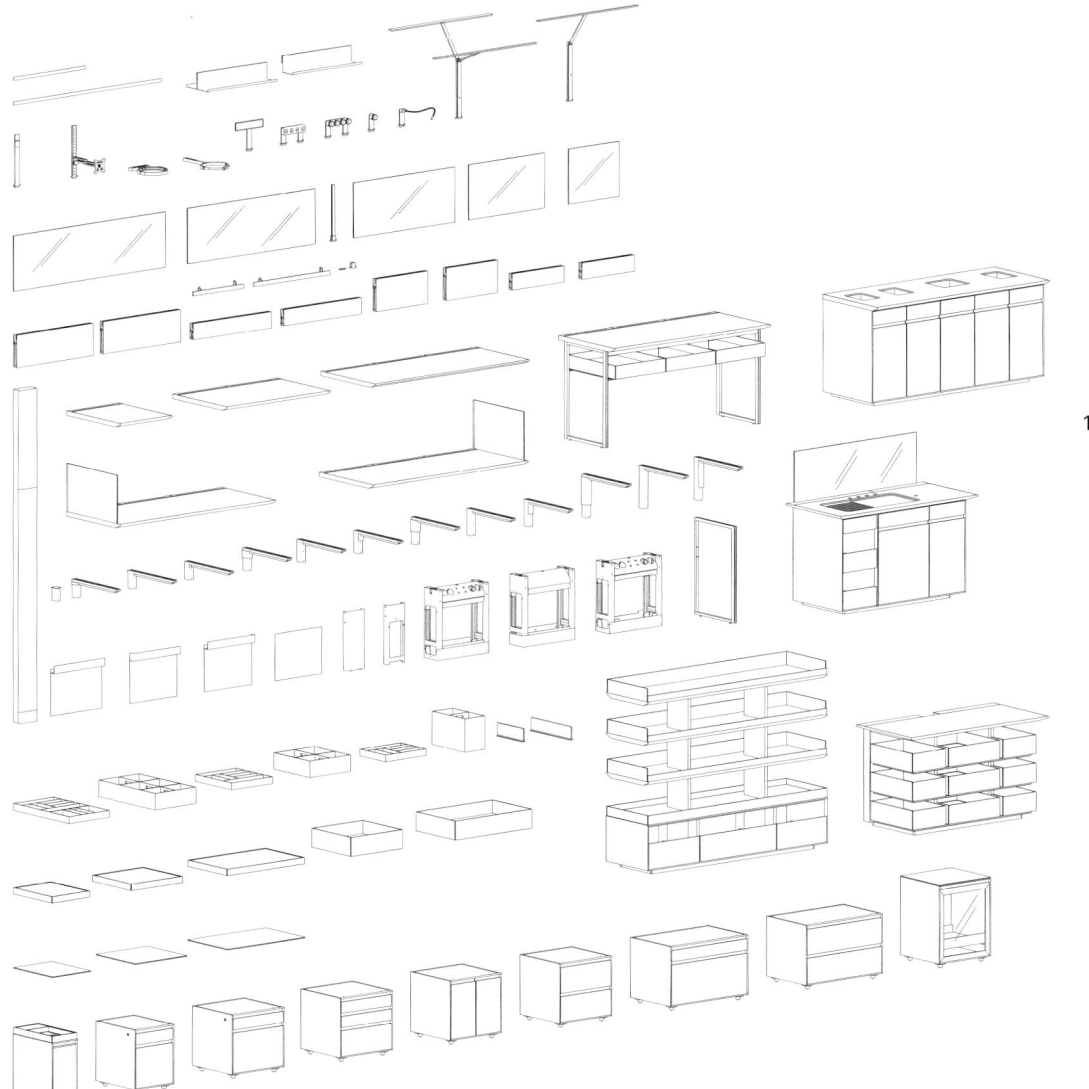

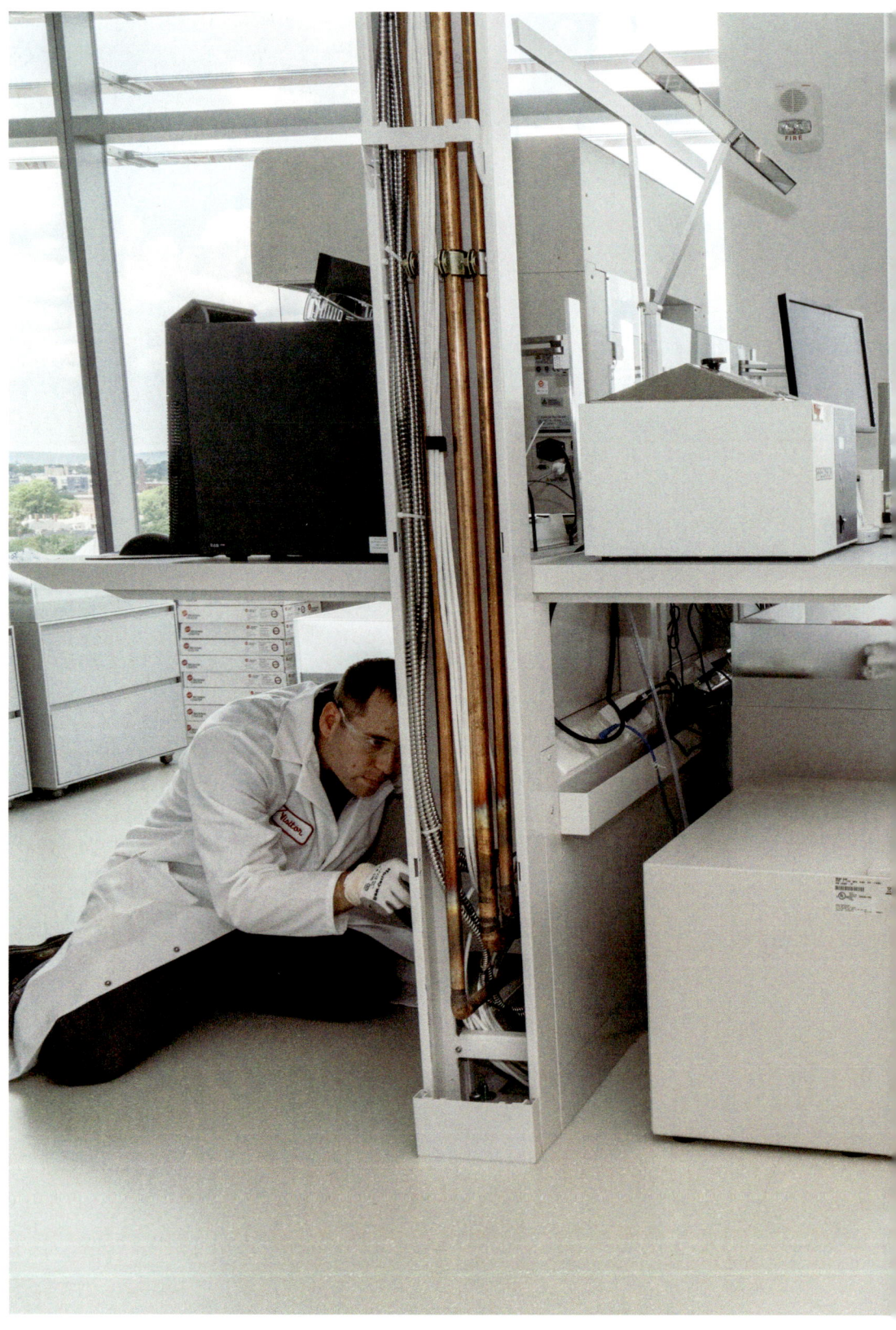

GENERALITY: ONE FOR ALL

The goal of NIBR, to discover new medicines, is
ultimately practical, but on a day-to-day basis there is
little to distinguish the activities in these labs
from those in basic science departments of academia.
Although the origins of the pharmaceutical industry
were in the chemistry of coal tar dyes, today chemistry
is only part of the armamentarium. Most space is
dedicated, as in biology labs of academia, to the discovery
of new biological pathways in health and disease
and to the analyis of protein and chemical structure at
the atomic level. In academic institutions as well as
industry, robots have relieved scientists of the need to
perform repetitive assays.

147

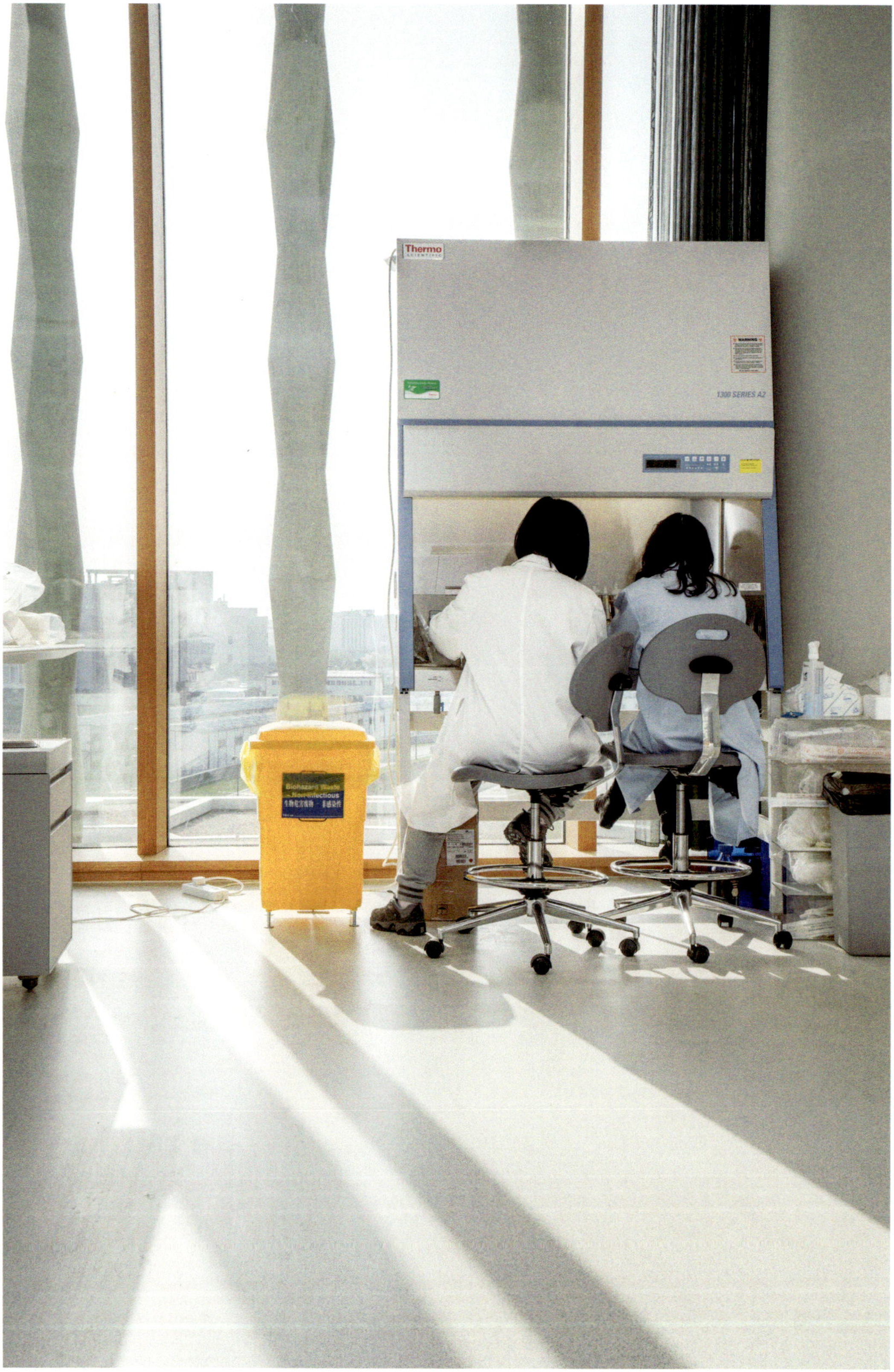
Thermo
1300 SERIES A2
Non-infectious

158

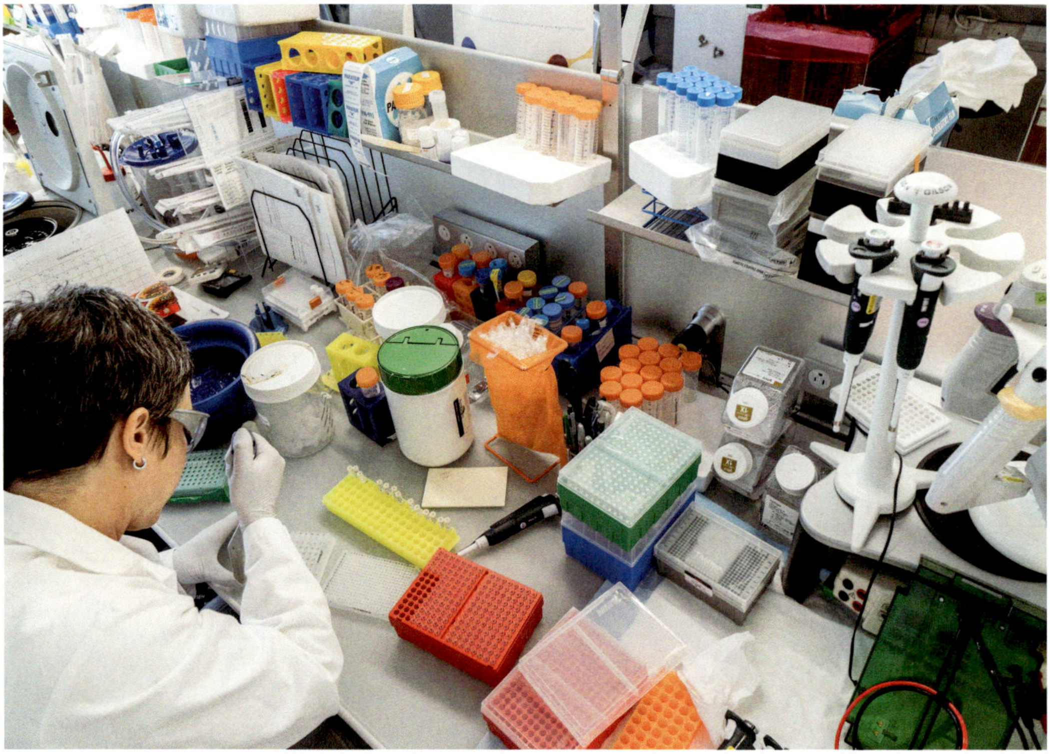

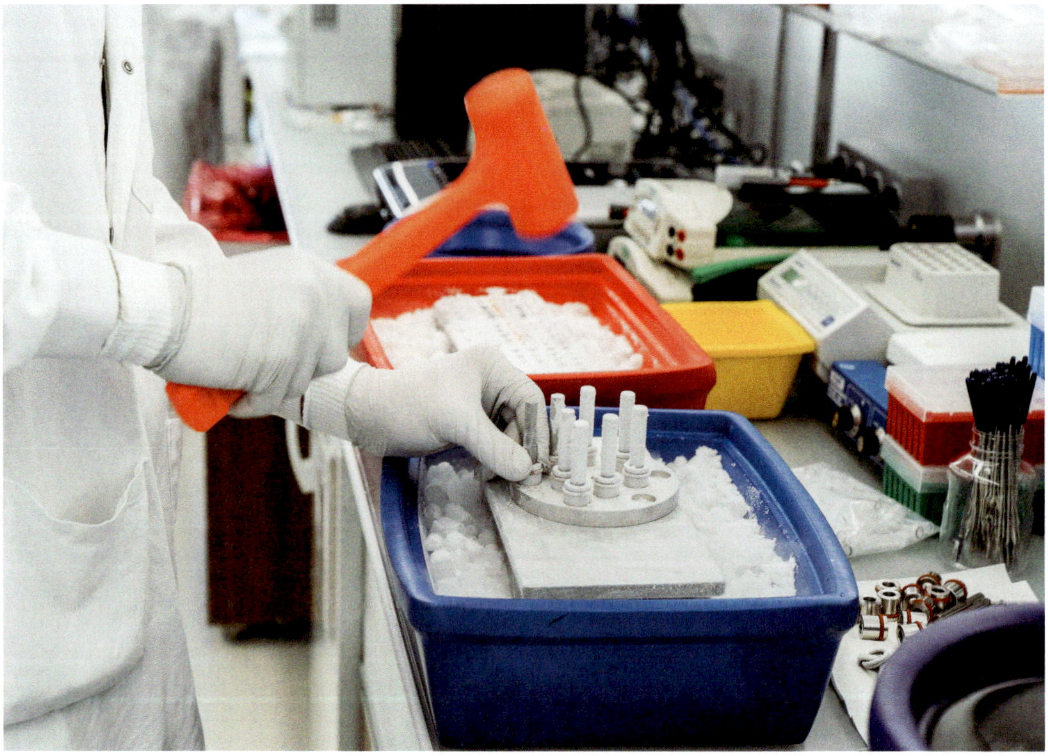

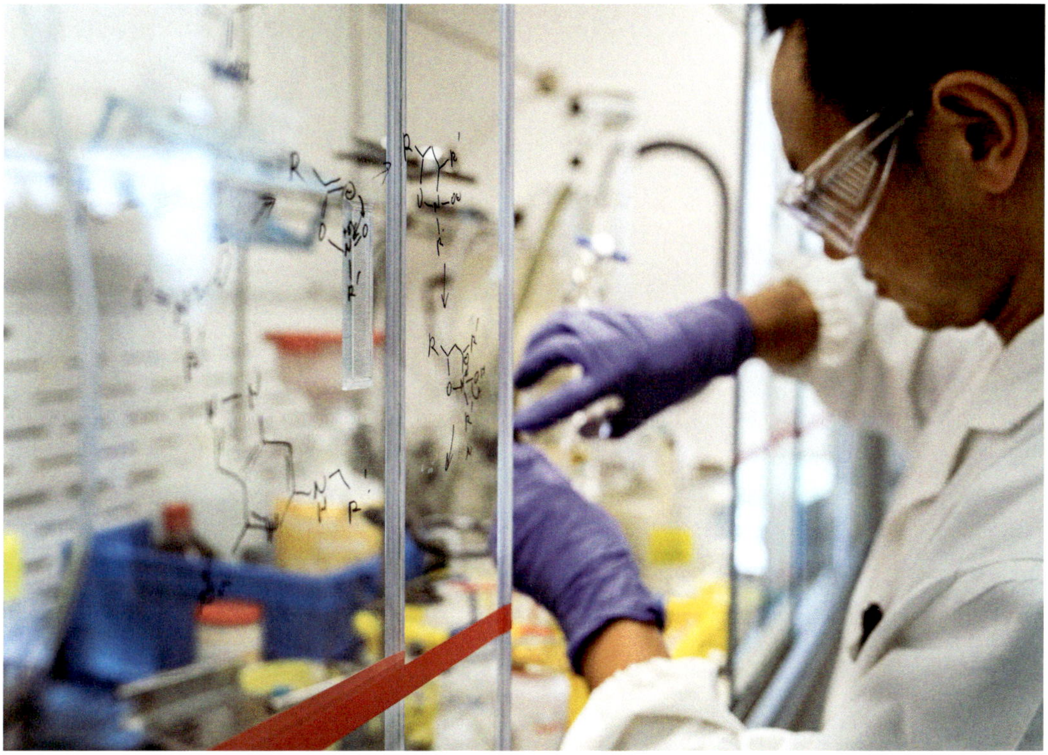

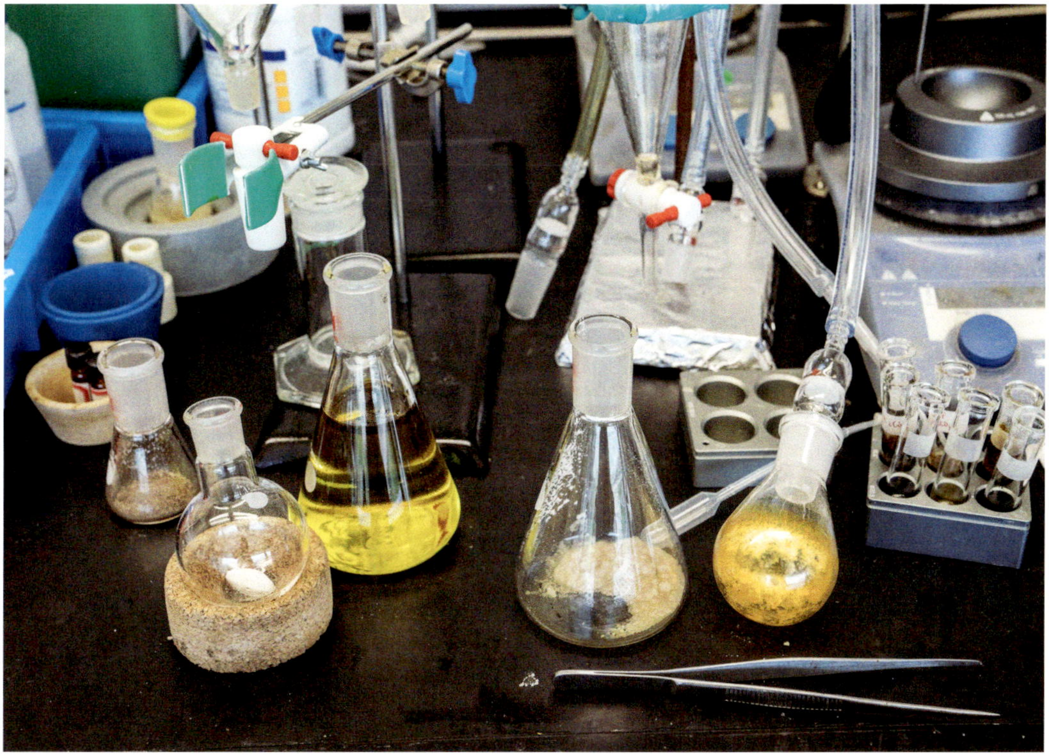

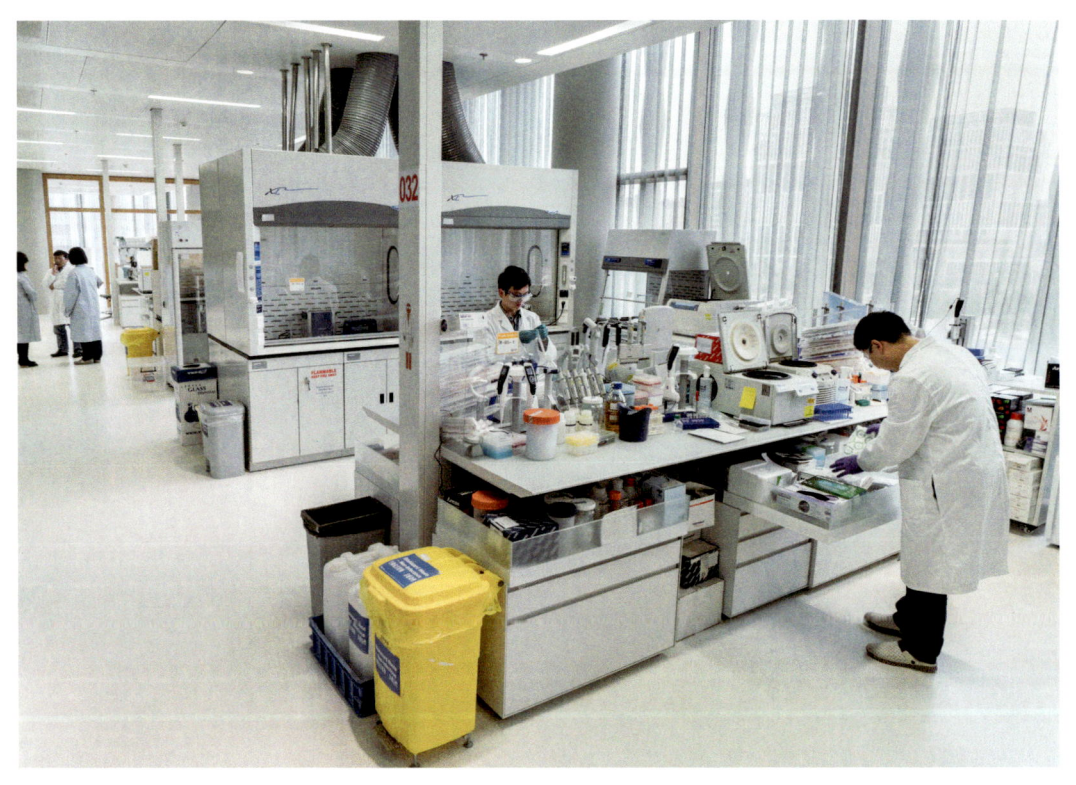

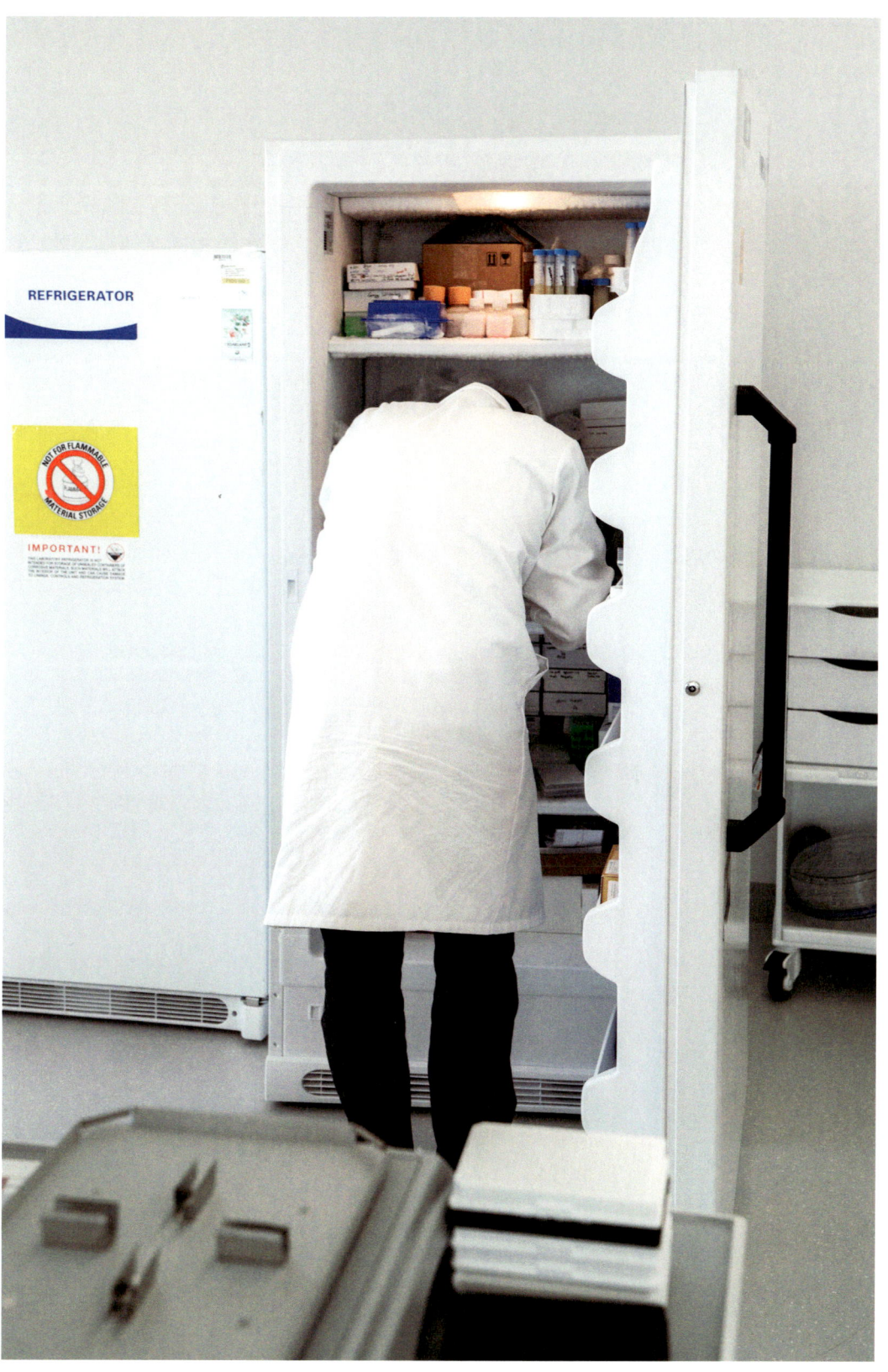

182

TWO CAMPUSES
CAMBRIDGE AND SHANGHAI

NIBR was established in 2002 as the scientific arm of Novartis, a multinational company based in Basel, Switzerland, by the merger of Sandoz and Ciba-Geigy, companies that had evolved on the banks of the Rhine from chemical to pharmaceutical companies. The goal of NIBR is to change the practice of medicine by discovery of new medicines based on patient need and fundamental science. The projects of NIBR span basic investigation, generation of medicines, and their testing in patients.

193

In some ways, the two sites chosen for new campuses could not have been more different: in Cambridge, Massachusetts, it is in the midst of an academic center, within walking distance of hundreds of biotech start-ups and major teaching hospitals; and in Shanghai in Pudong, away from the center of town, on an island in an industrial park, surrounded by a small moat. But the goals are identical.

CAMBRIDGE

The Cambridge campus serves as the international headquarters for NIBR. Cambridge has long been home to many biotechnology companies, but Novartis was the first large pharmaceutical company to move a major proportion of research to the area, and the only one to make the site its global headquarters. Many other pharmaceutical companies later established branches in the area, which has become a hub for such activities.

MIT and biotechnology companies surround the NIBR site. We have embedded our campus in the community, making the central garden accessible to neighbors, providing space for retail stores within our buildings along the street, and including a laboratory for the teaching of science to Cambridge middle school students.

The meandering paths and small picnic tables in the central courtyard extend the open feeling of the interior design. Because of the angle of views and landscaping, from inside the campus feels closed and protected, while from the streets it looks open and accessible. The New England farm feeling is reinforced by the locally sourced Chelmsford granite for the exterior of the Maya Lin building, and the native gray birch, American beech, and rosebay rhododendron in the courtyard.

The campus includes two new buildings and a smaller brick building, not renovated, which housed (from 1944 to 1959) the Whirlwind computer of MIT, the first real-time high-speed computer using random-access magnetic core memory.

Construction was complicated by the superficiality of the water table, requiring buttresses to hold up the basement walls as concrete was poured.

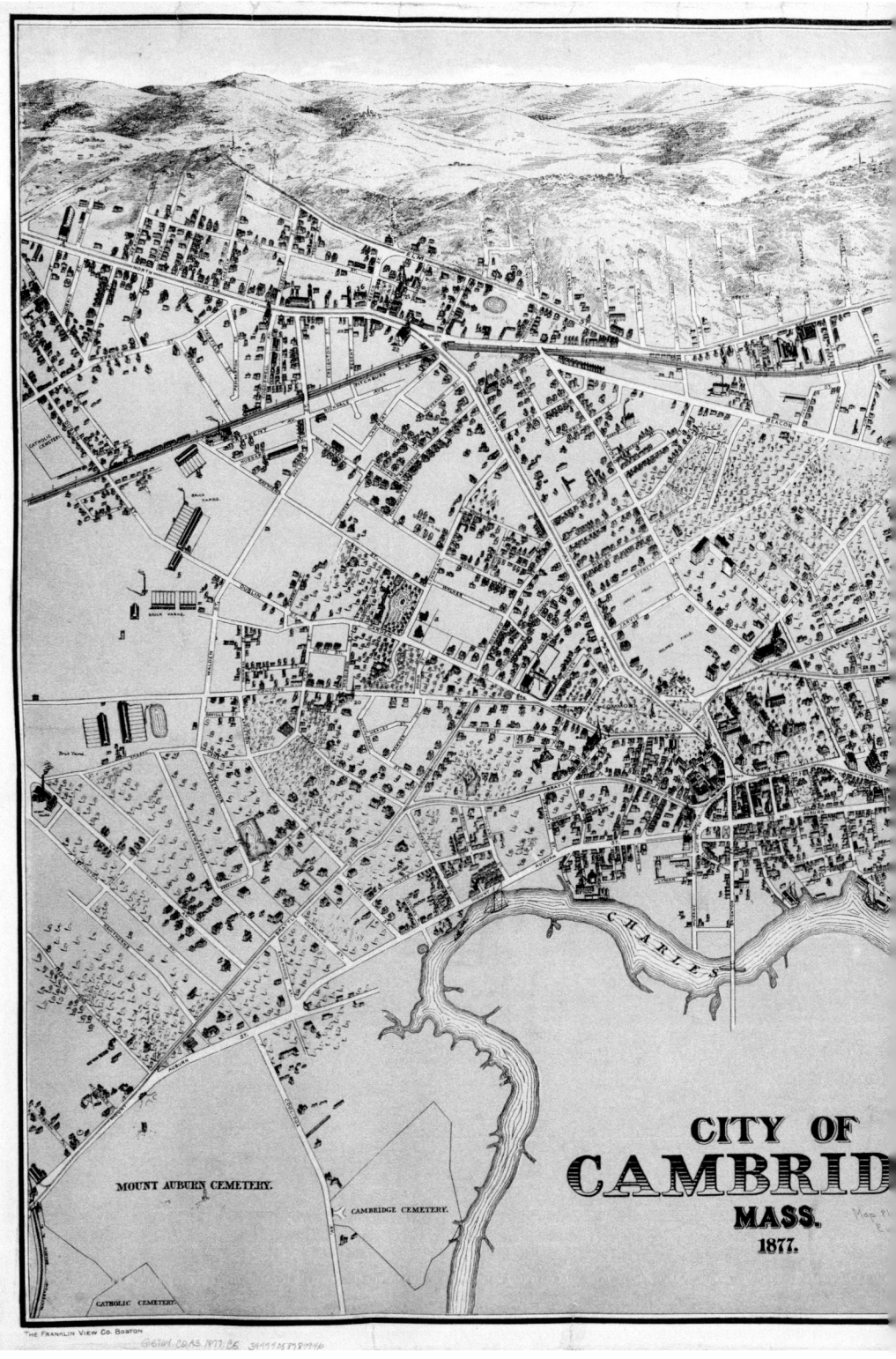

Red dot shows approximate location of the campus

200

181 Massachusetts Avenue

22 Windsor Street

0 5 10   30 m

SITE MASTER PLAN
MAYA LIN STUDIO

As site master planner, Maya Lin was faced with the challenge of resolving the clash of three distinct city grids dating from the 1850s, 1870s, and 1890s. Using form, massing, and careful material selection, all three buildings successfully mediate the conflicting directions enforced by the historic grids. Taken together, the three buildings that occupy the site dialogue with one another and with surrounding buildings that define the neighborhood. By using a human scale on the periphery of the site closest to foot traffic, the arrangement of buildings acts as an edge that defines and protects a university campus-like outdoor environment meant to celebrate the intersection of Cambridge's civic life and the Novartis scientific community. In keeping the range of materials used in the buildings simple and few in number—earth, stone, and glass—a layered effect is achieved that serves to connect the exterior of the new campus with its surroundings, while high-lighting and showcasing the activities taking place within the buildings, creating a glowing beacon to the community.

Designer and Master Planner	Maya Lin Studio with Bialosky and Partners New York, NY, USA	**Floor area above ground**	30,658 m² 330,000 sq. ft.
Executive Architect	Cannon Design	**Floor levels above ground**	8 (science tower); 4 (non-lab wing)
		Building height above ground	45 m 148 ft., 6 in.
		Workplace capacity	Approximately 500–560 (lab and non-lab)
		Materials and structure	Steel and concrete structure, stone, and glass

Maya Lin, born 1959, received her master's degree in architecture from Yale University in 1986 and established her studio in 1991. Having virtually redefined the idea of the monument with her very first work, the Vietnam Veterans Memorial, Lin has gone on to pursue a remarkable body of work in both architecture and art. Projects range in scale from private residences to nonprofit institutions.

The studio design ethos is to create a close dialogue between the landscape and built environment, and is always committed to advocating sustainable design solutions and practices. Lin has been the subject of numerous solo art exhibitions at museums and galleries worldwide. She has also created permanent outdoor installations for public and private collections around the world.

206

181 Massachusetts Avenue, L4

0   5   10   20 m

MAYA LIN

181 MASSACHUSETTS AVENUE

Building on the themes established in the master
plan, 181 Massachusetts Avenue designed by Maya Lin
Studio is really two buildings in one: a low-rise
along Massachusetts Avenue, and a tower set back from
the edge of the site. When examined closely, it becomes
apparent that the tower emerges from the low-rise section
of the building as a gesture of respect to the human
scale of the neighborhood; the modest height also permits
sunlight into the courtyard. Materially, Maya took
her ornamental façade inspiration from the stone walls
commonly found woven into the fabric of the New England
landscape. Made from locally sourced Chelmsford
granite and influenced by the microscopic structure of
bone, the ornamental façade contains a matrix of openings
that allow transmission of light to interior occupants.
The light and airy pattern established by the ornamental
façade is further abstracted and carried through to
the tower where back-painted glass, a variety of frits,
and randomly dispersed clear vision panels give the
glass façade a three-dimensional and variegated appearance as the sun angle changes throughout the day.

212

214

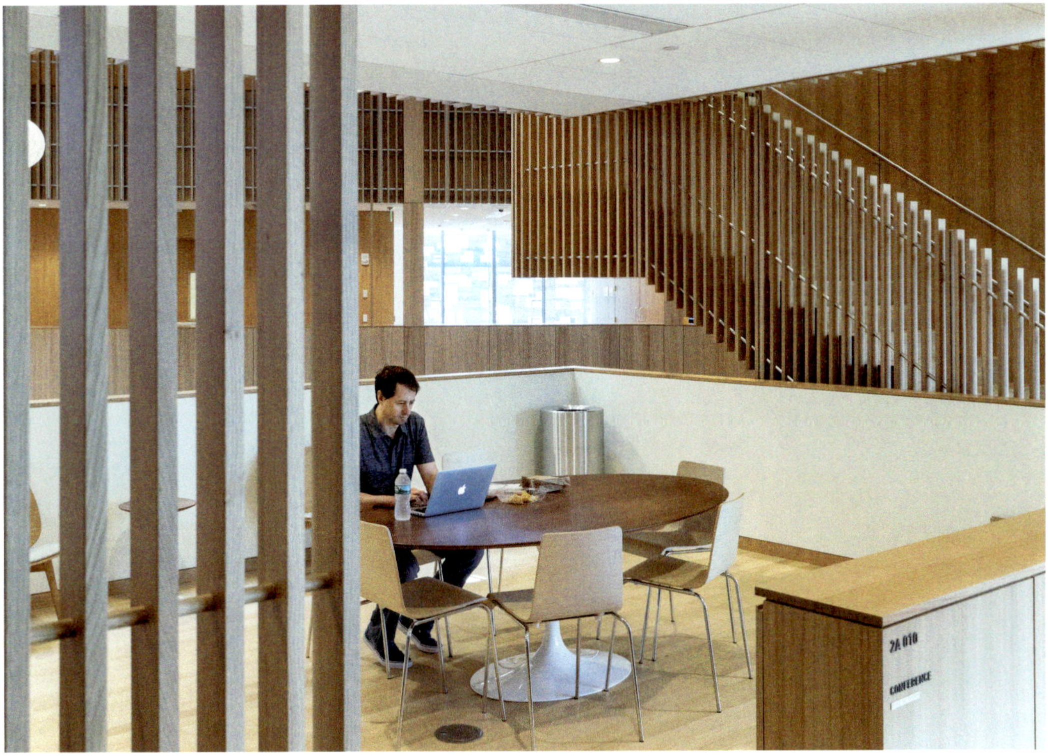
2A 010
CONFERENCE

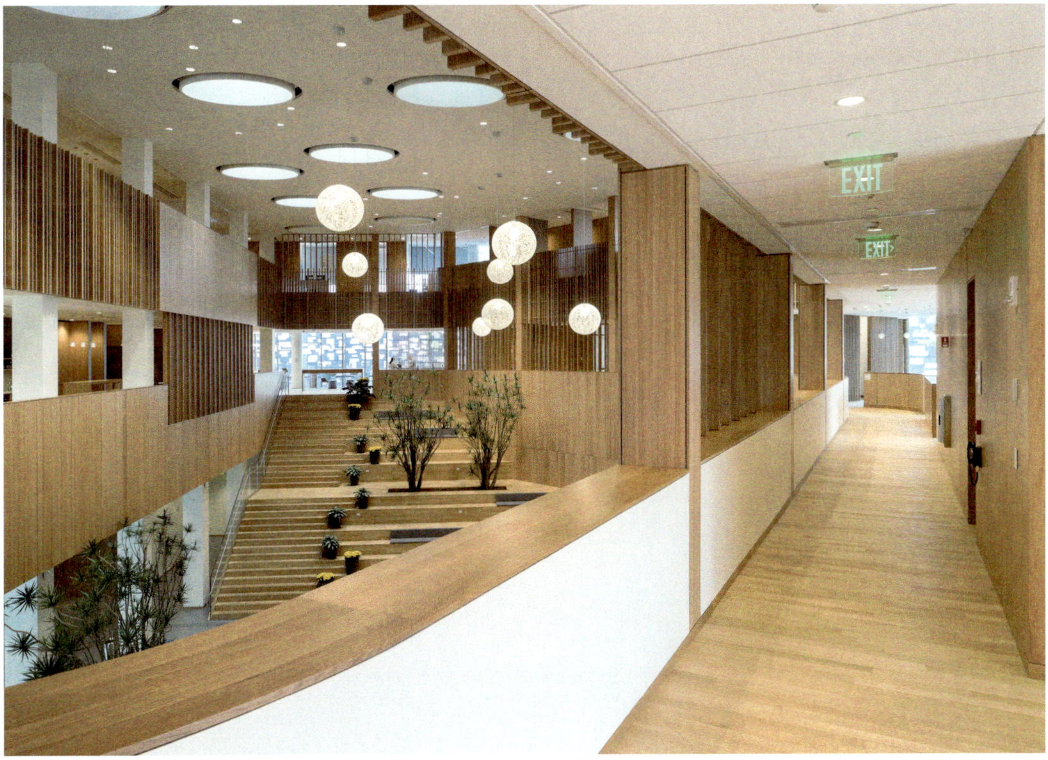
EXIT
EXIT

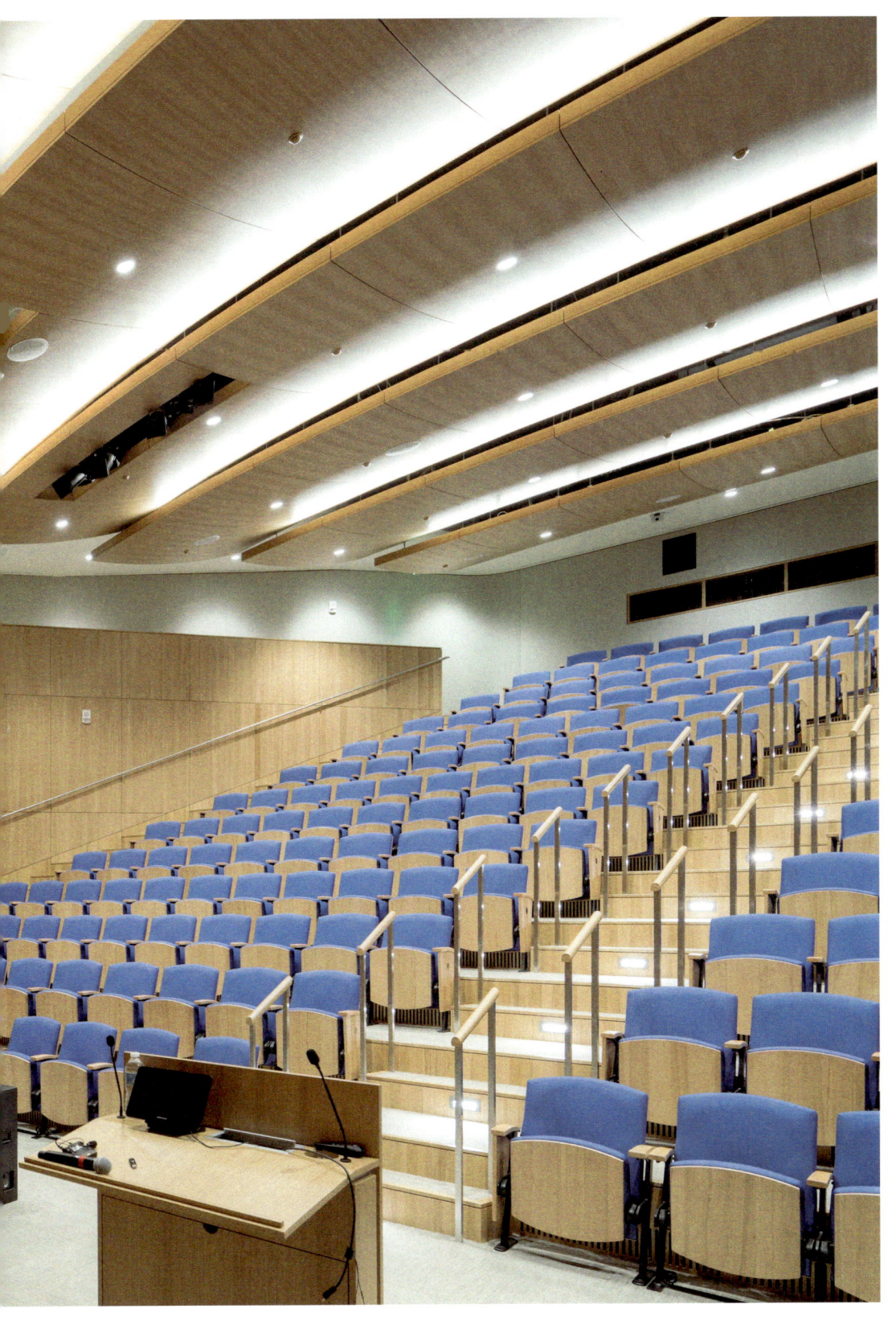

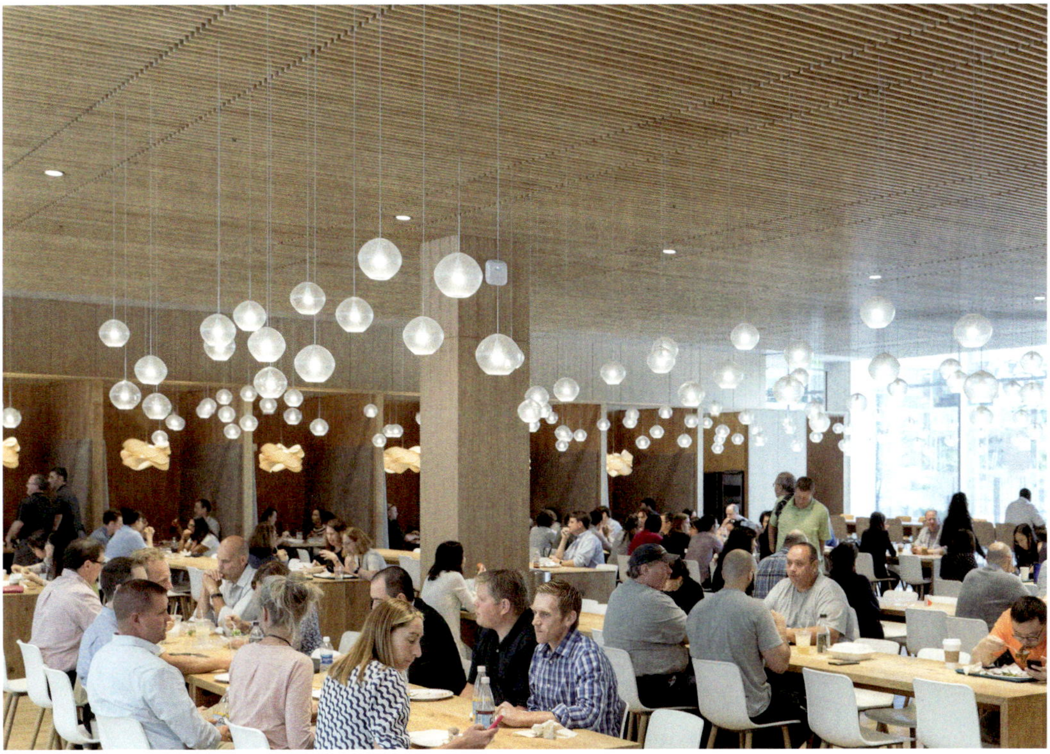

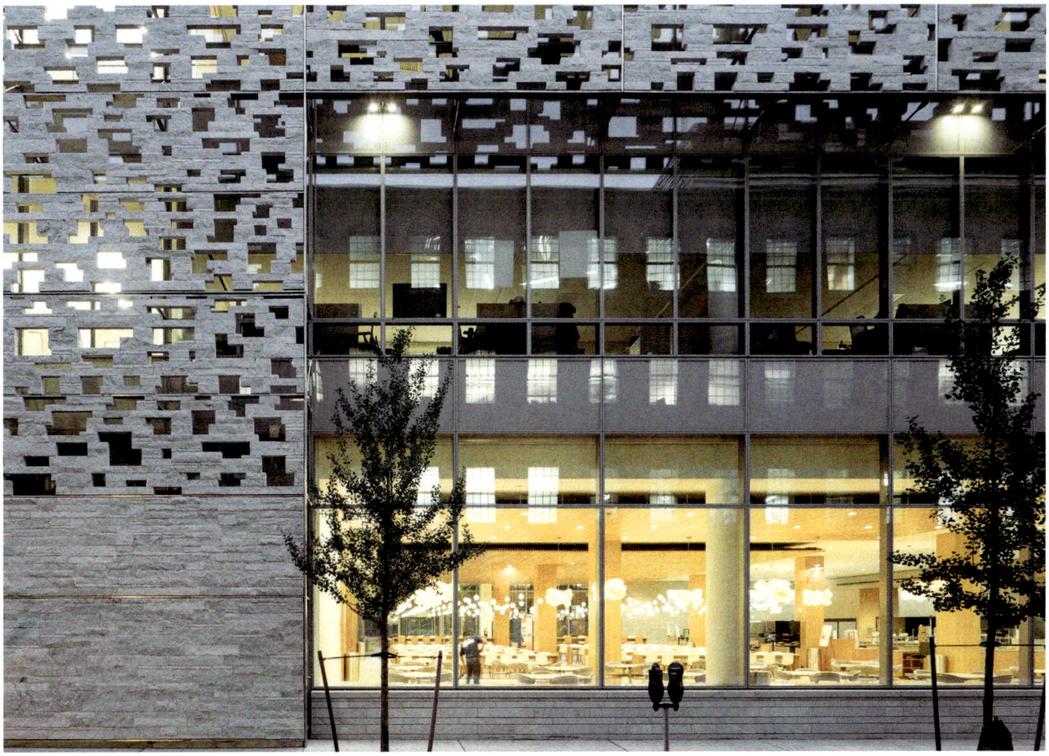

Design Architect	Toshiko Mori Architect PLLC New York, NY, USA	Floor area above ground	26,477 m² 285,000 sq. ft.
Architect of Record	Cannon Design	Floor levels above ground	7 (all laboratory)
		Building height above ground	46 m 151 ft., 3 in.
		Workplace capacity	520–550
		Materials and structure	Steel and concrete structure, stone, and glass

Toshiko Mori, born 1951, is the Robert P. Hubbard Professor in the Practice of Architecture at Harvard University's Graduate School of Design and the principal of Toshiko Mori Architect, founded in 1981. The practice is known for its innovative and influential work in a diverse body of projects. Mori's intelligent approach to ecologically sensitive siting strategies, historical context, and innovative use of materials reflects a creative integration of design and technology.

Her designs demonstrate a thoughtful sensitivity to detail and involve extensive research into the site conditions and surrounding context. The work of TMA combines a strong conceptual and theoretical approach with a thorough study of programmatic needs and practical conditions to achieve a design that is both spatially compelling and pragmatically responsive. Toshiko Mori Architect continues to engage in an architecture of material exploration, technological invention, and theoretical provocation.

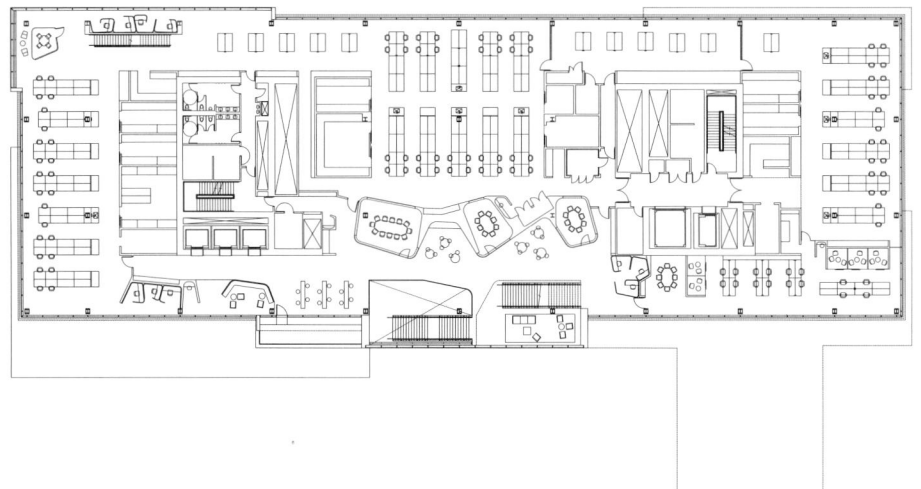

22 Windsor Street, L4

0 5 10 20 m

TOSHIKO MORI

22 WINDSOR STREET

Positioned along the north end of the central courtyard, the building designed by Toshiko Mori Architect PLLC presents the broad side of the structure to Massachusetts Avenue, effectively making it the face of NIBR. To prevent the size of the building from overwhelming the courtyard, overhangs and volumetric shifts were used to break up the geometry of the façade. The impact of this approach is especially apparent as the ceremonial staircase on the interior of the space manifests itself on the exterior as double-height "mini atria"—each with its own outdoor balcony. In keeping with the master planner's concept, warm-toned Jura stone was used to complement the granite and cool tones used in Maya Lin's façade. Clear vision panels are used throughout the building to express and reveal the scientific lab as an identifying feature, and to present the humanity of NIBR to the Cambridge community as a "glowing beacon" at the center of the campus. Louvers made from copper mesh inserted between planks of clear glass provide relief from the high sun angles in the summer, while allowing low-angle sun in the winter to penetrate deep into the lab spaces— at high sun angles in particular, the acrylic planks behave as a prism, refracting sunlight in subtle, multi-colored bands.

230

232

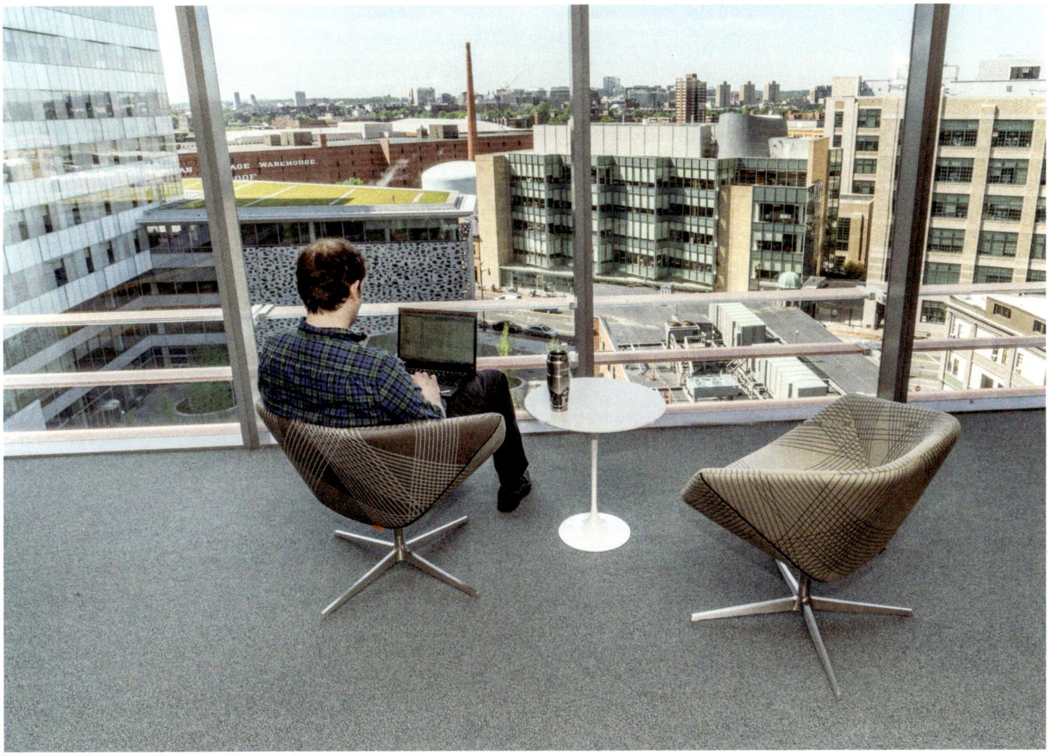

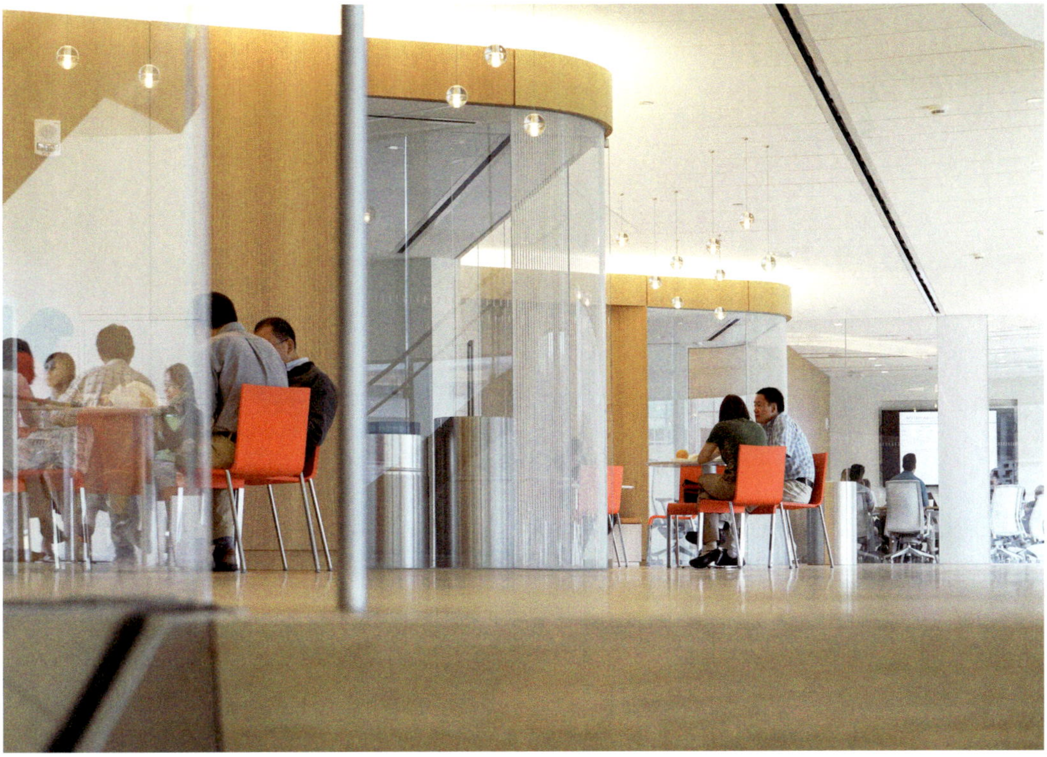

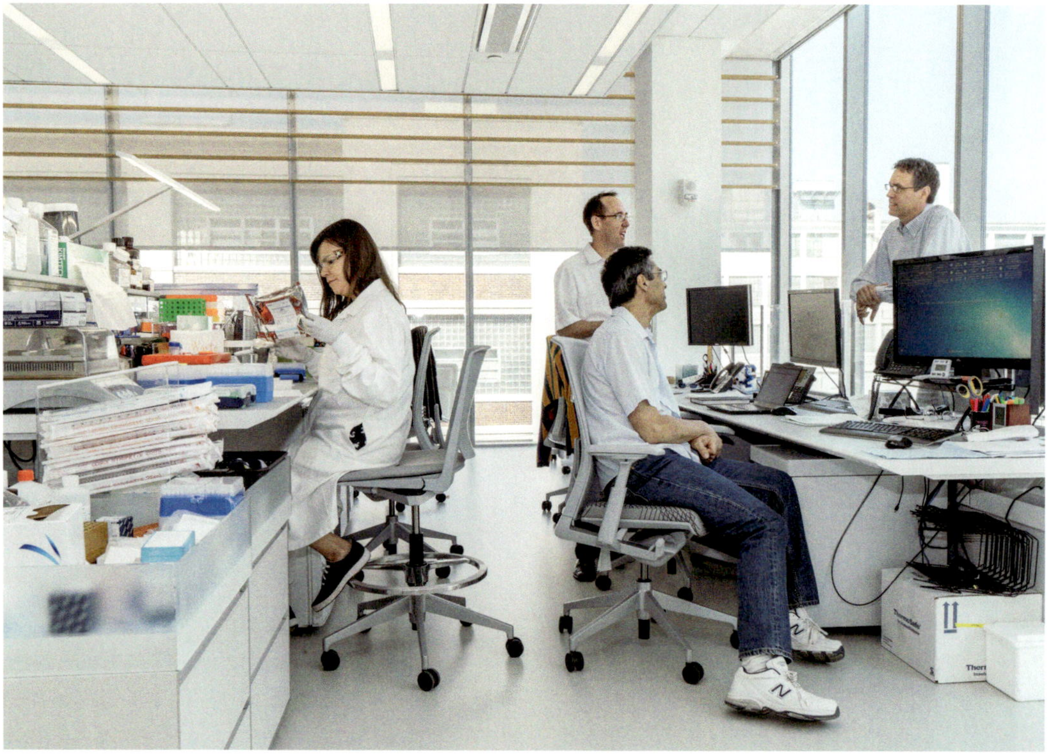

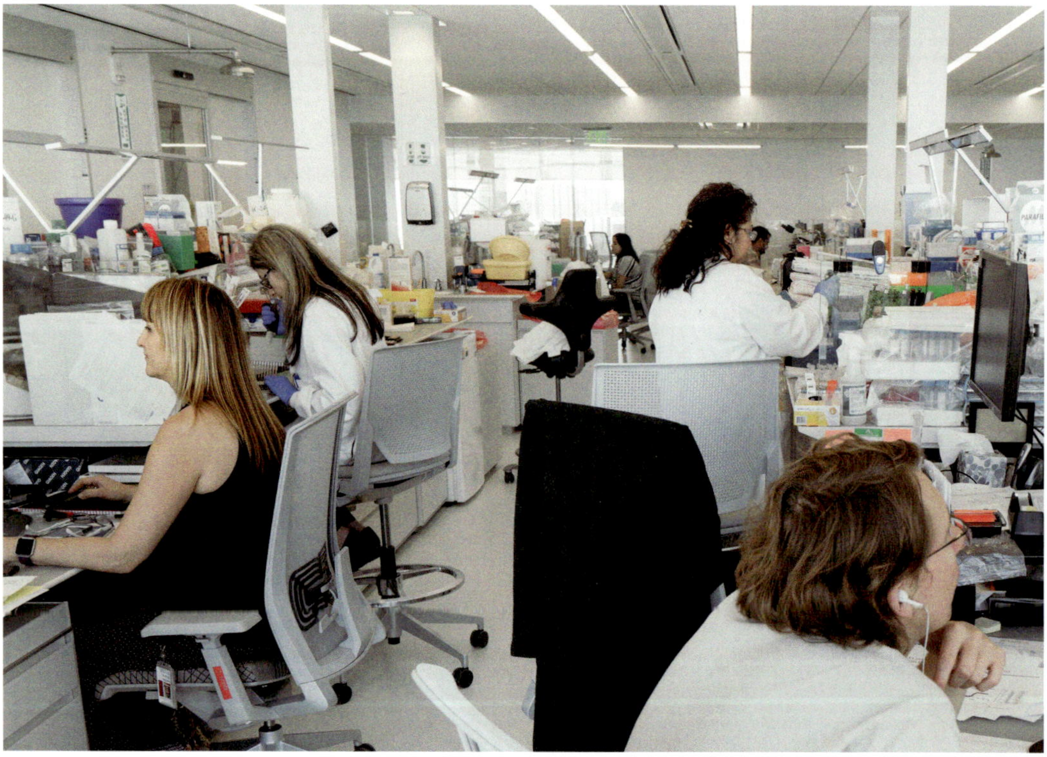

Architect   Michael Van Valkenburgh Associates
Cambridge, MA, USA

Michael Van Valkenburgh, born 1951, is the Charles Eliot Professor in Practice of Landscape Architecture at Harvard University's Graduate School of Design. In 1982 he founded Michael Van Valkenburgh Associates in Cambridge, Massachusetts, USA. MVVA is inspired by the power of landscape architecture to deliver beauty in its many forms: rational, lyrical, and exuberant. They believe that the most meaningful landscapes anticipate what they will signify to their many audiences and emerge from straightforward, elegant problem-solving techniques. MVVA's work is unfussy and original, sustainable and enjoyable, democratic and direct. They believe that landscape plays a critical role in shaping a positive experience of place—both found and designed—and that a strong sense of place helps enhance everyday experience, healthy communities, and quality of life, which is at the heart of a livable city. Their work encompasses a range of scales and landscape types, from large public parks and university campuses to intimate gardens and private landscapes.

MICHAEL VAN VALKENBURGH

LANDSCAPE

At the heart of the Novartis Institutes for Biomedical
Research campus is a courtyard built atop a subterranean
parking structure. It is a crossroads both for NIBR staff
and the larger public, providing pedestrian connections,
open green space, and a landscape of momentary refuge
from the stresses of urban and professional life. A gener-
ously proportioned entry plaza on Massachusetts
Avenue marks an intermediate hub of pedestrian activity
between Kendall and Central Squares, while on the
campus interior, meandering walkways connect the NIBR
buildings to smaller-scale outdoor rooms with moveable
chairs and tables. The ground plane slopes gently down
toward Massachusetts Avenue, providing views both
into the campus from the street and out onto the city
from within the campus. An absorbing grove of trees,
composed of a mix of deciduous, evergreen, and flowering
species marks changes in the seasons and helps to unify
the campus' heterogeneous architecture. A mix of
low evergreen species and herbaceous perennials creates
a continuous groundcover punctuated by occasional
clusters of shrubs that strategically contain smaller
spaces and screen views.

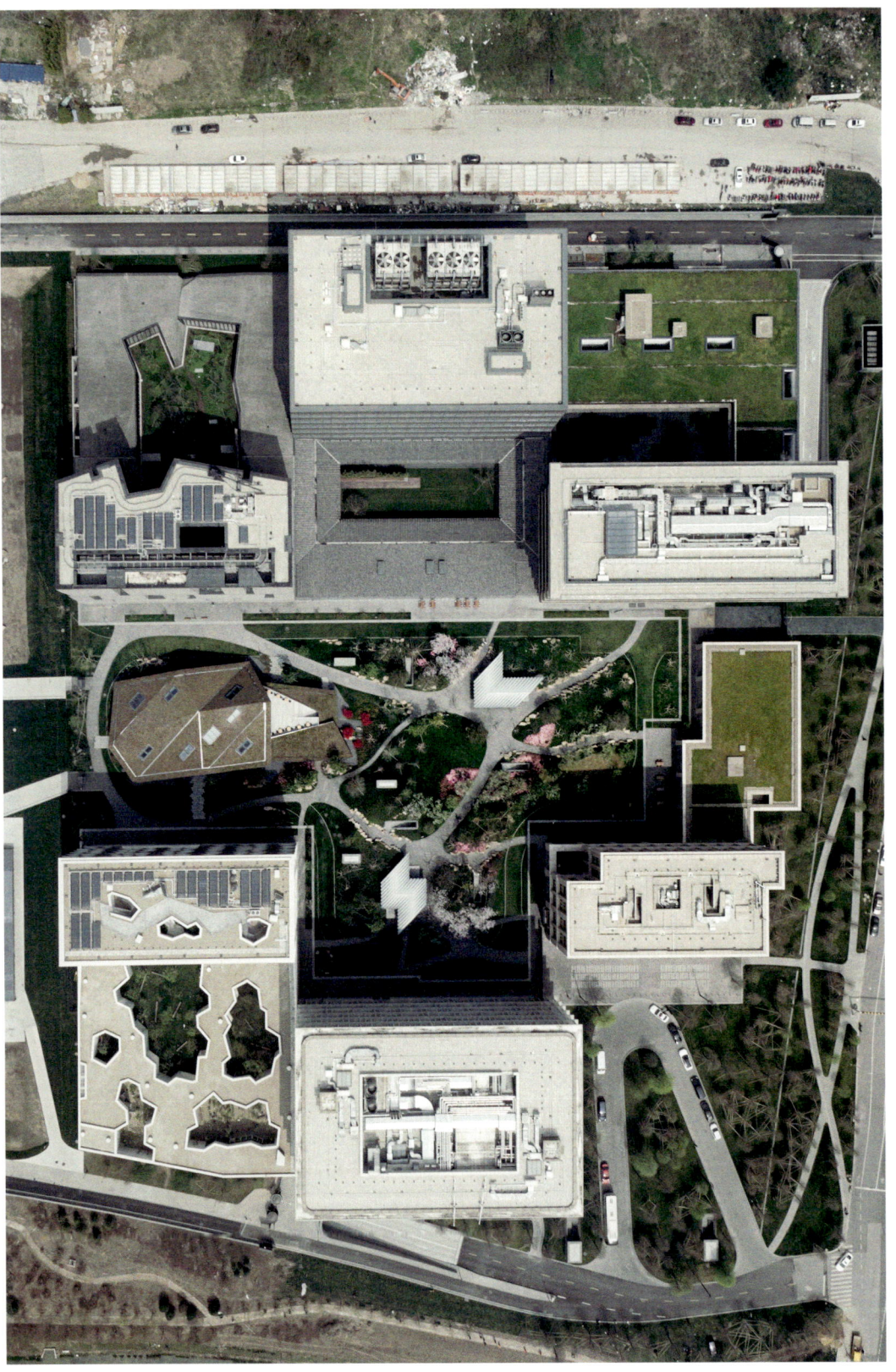

SHANGHAI

In Shanghai we had the opportunity to build on an island separated from the rest of the Zhangjiang Biotechnology Park by a canal. (Like in most of Shanghai, this means basements are below water level and need special attention to waterproofing and buttressing.) The original hope had been to have a series of two-story buildings designed for science but resembling teahouses in style, surrounded by classical Chinese gardens, designed for contemplation by attention to shade and light, placement of stones, meandering paths, gentle streams, pavilions, and bridges. For a variety of reasons, we had to increase the height and overall size of the buildings, so the site plan changed to incorporate small gardens within each building and more formal landscaping in the center of the campus. Most of the new buildings remained totally dedicated to science, and others became office buildings.

There is no overarching theme to the architecture of the biotechnology park, as, in fact, there is none to architecture in Shanghai in general. The ornate granite and marble neoclassical buildings along the Bund overlooking the Huangpu River were built by English, French, and Americans in the late 1800s and early 1900s, many of them tied to the opium trafficking trade. This Western influence evaporated after the Japanese invasion in December 1941. Modern buildings in Shanghai are notable for their scale and quirkiness of design. The buildings of our campus integrate design principles and materials reflecting traditional Chinese and Western influences.

248

Shanghai City (1927)
Red dot shows approximate location of the campus

Middle I.
Chungpao Sha
Block House I.
PAOSHAN CITY
WUSUNG
CHINESE CITY
Choupu
Zauwukyung
Nekokahang
Scale 1:50,000

Architect   Atelier FCJZ
                       Beijing, China

Project architect Liu Xiang Hui, Keith Wu,
                       Simon Lee

250

SITE MASTER PLAN
Yung Ho Chang, Atelier FCJZ

The site master plan creates multifunctional, flexible, and innovative workplaces that meet international standards while remaining sensitive to the context of China. The model of ancient Chinese spatial planning is the basic idea for the site master plan, while each building embraces a closed courtyard and is geometrically aligned around a central plaza area. Following the concept of a Chinese teahouse, the cafeteria is placed in the middle of the central plaza area.

All workplaces are designed to be interactive and collaborative, bringing together different disciplines, while providing domains for escape, privacy, and meditation.

254

Monks at groundbreaking

Architects	Sergison Bates Architects London, Great Britain	**Floor area above ground**	6,900 m² 74,270 sq. ft.
Project architect	Jerry van Veldhuizen	**Floor levels above ground**	6
		Building height above ground	32 m 105 ft.
		Workplace capacity	96
		Materials and structure	Terracotta tiles, brass façade panels

Jonathan Sergison, born 1964, graduated from the Architectural Association School of Architecture in London in 1989. Since 2008 he has been a professor of design and construction at the Accademia di Architettura in Mendrisio, Switzerland.

Stephen Bates, born 1964, graduated from the Royal College of Art in London in 1989. Since 2009 he has been a professor of urbanism and housing at the Technische Universität in Munich.

In 1996 they established Sergison Bates Architects and have earned a reputation as one of the UK's leading architectural practices by successfully engaging with all dimensions of architectural and urban design. International in outlook and staff composition, they opened a second studio in Zurich in 2010 and are currently involved in a wide range of international projects ranging from urban planning to regeneration, public buildings, and housing.

Sergison Bates' projects are informed by a sensitive approach to place, the experiential potential of materials and construction, and a concern for the environmental, social, and economic aspects of sustainability. They aspire to create an architecture that is contemporary and rooted in its context, at all scales, and are committed to a research-based approach, supported by the partners' academic work.

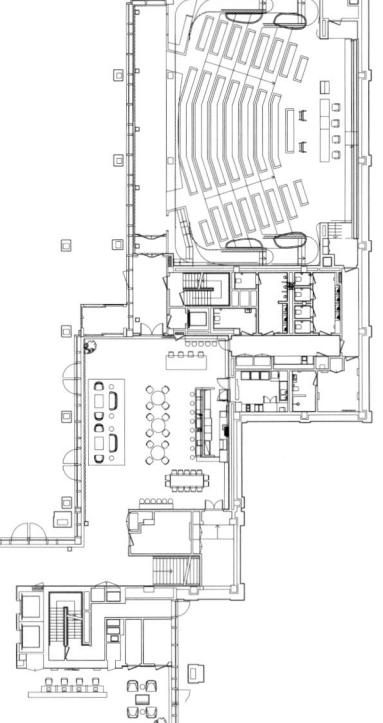

Building 1, L1

0   5   10   20 m

SERGISON BATES ARCHITECTS

BUILDING 1

The office building serves as the campus entrance building and visitor welcome center. Its formal appearance makes reference to the classical architecture of the Shanghai Bund from the nineteenth century. The dominant terracotta columns in the façade give the building a vertical expression. Horizontal stone registers and deep recessed infills with brass panels further articulate the classical elegance of this building's façade.

The interior is open and welcoming and acts as the host to the whole Novartis Shanghai campus, its community and guests. A variety of public and private spaces are the essence of this building. Multiple but simple daily acts are framed in a range of atmospheric spaces that bring a sense of care and ease. It is a place for people to find a favorable corner to talk, to slow down, with spaces that are comfortable for reflection, pause, and connection. Visitors to this building are encouraged to mix and mingle, living the Novartis culture of drawing people together.

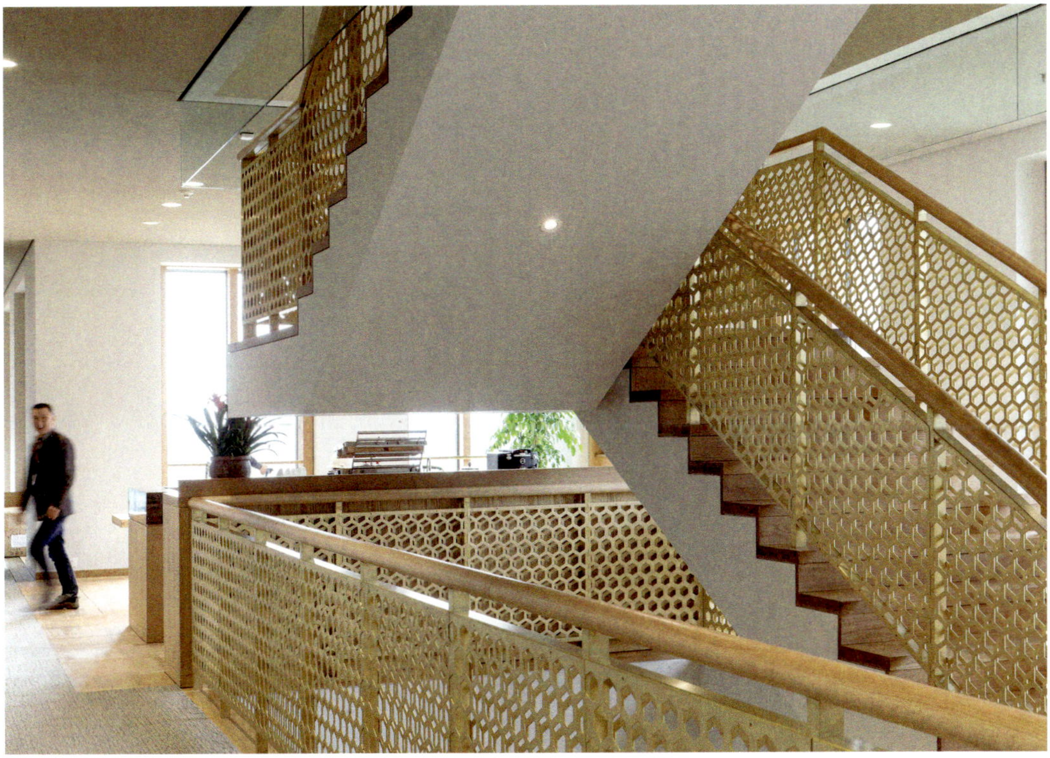

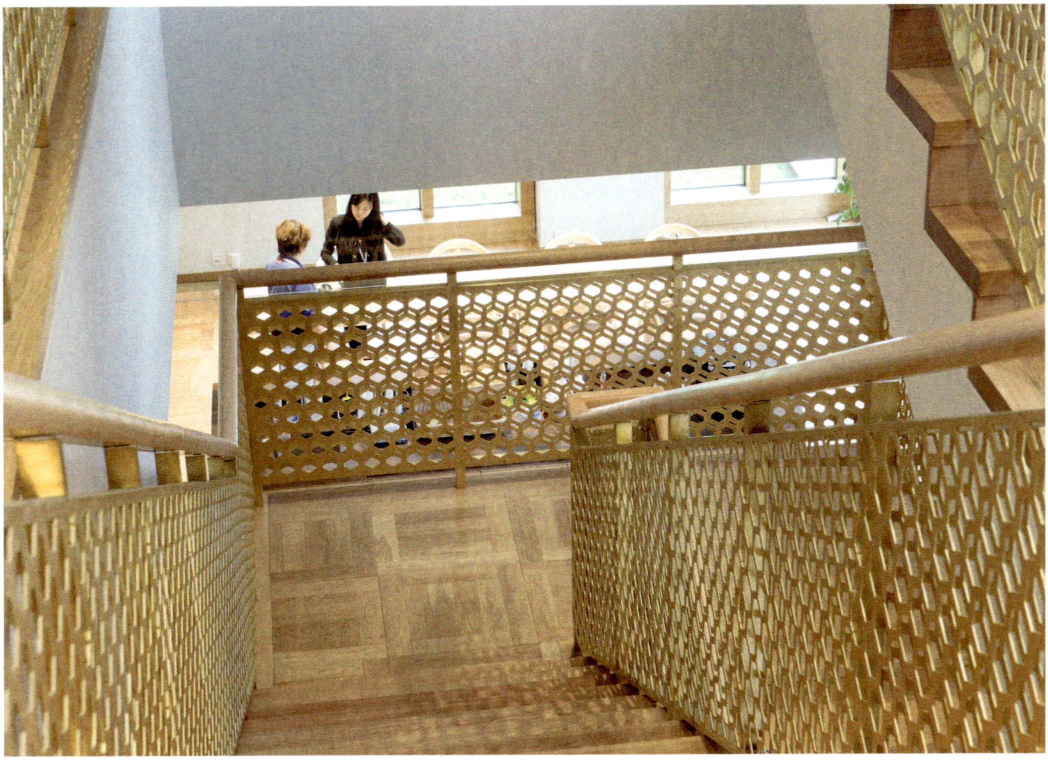

Architects	Jiakun Architects	Floor area	9,219 m²
	Sichuan, China	above ground	99,232 sq. ft.
Project architect	Liu Jiakun	Floor levels	6
		above ground	
		Building height	32 m
		above ground	105 ft.
		Workplace capacity	218
		Materials and	Bamboo elements,
		structure	recycled brick façade

Liu Jiakun, born 1956, founded Jiakun Architects in 1999. As a multidisciplinary office, Jiakun Architects is specialized in architectural design as well as large-scale planning, urban design, landscape design, interior design, product design, and installation art. The office has organized and participated in multiple international collaborations and exhibitions.

Their clients are widely distributed over domestic and overseas markets. Their designs address contemporary architectural issues with a sense of realism—an approach inspired by folk wisdom. With faith in the compatibility of tradition and reality, Jiakun Architects has devoted itself to translating the cultural essence of the Orient into contemporary architectural language.

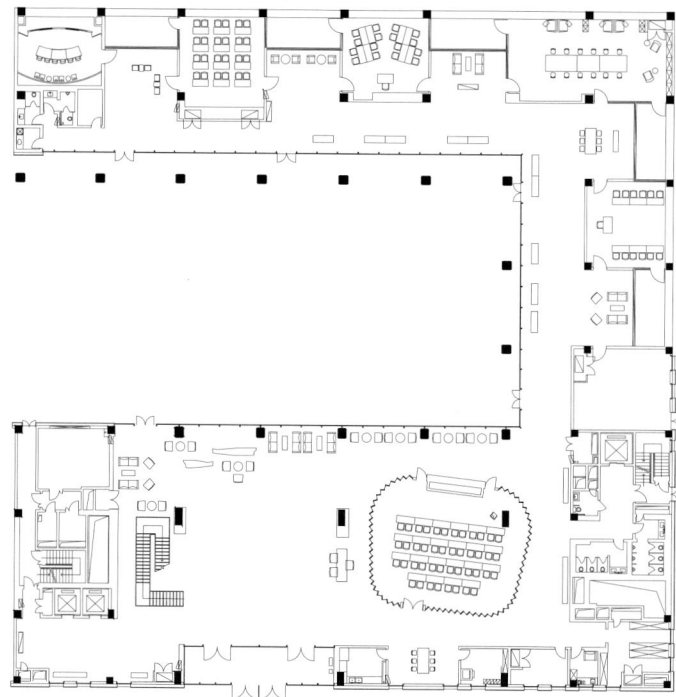

Building 2, L1

0   5   10   20 m

LIU JIAKUN

BUILDING 2

The office building is composed in a traditional Chinese context with a base, body, and roof. A particular emphasis has been given to the cantilevered expression of the roof eaves and the horizontal extension of the floor slabs. These accessible balconies invite employees to step out for a moment and enjoy the outdoors from an elevated level, protected by eyelash-like shading elements. Recycled bricks build up a solid base and overall façade impression.

The interior design is based on the overall campus workspace concept, providing open-plan offices, meeting rooms, print/copy facilities, and break areas. A calm and homelike interior welcomes employees. Carpets and light-colored bamboo frame the various areas. Classical Chinese furniture invites people to take a seat in the public and leisure areas. As a gentle contrast, contemporary furniture defines the office spaces, coffee points, and the academic area.

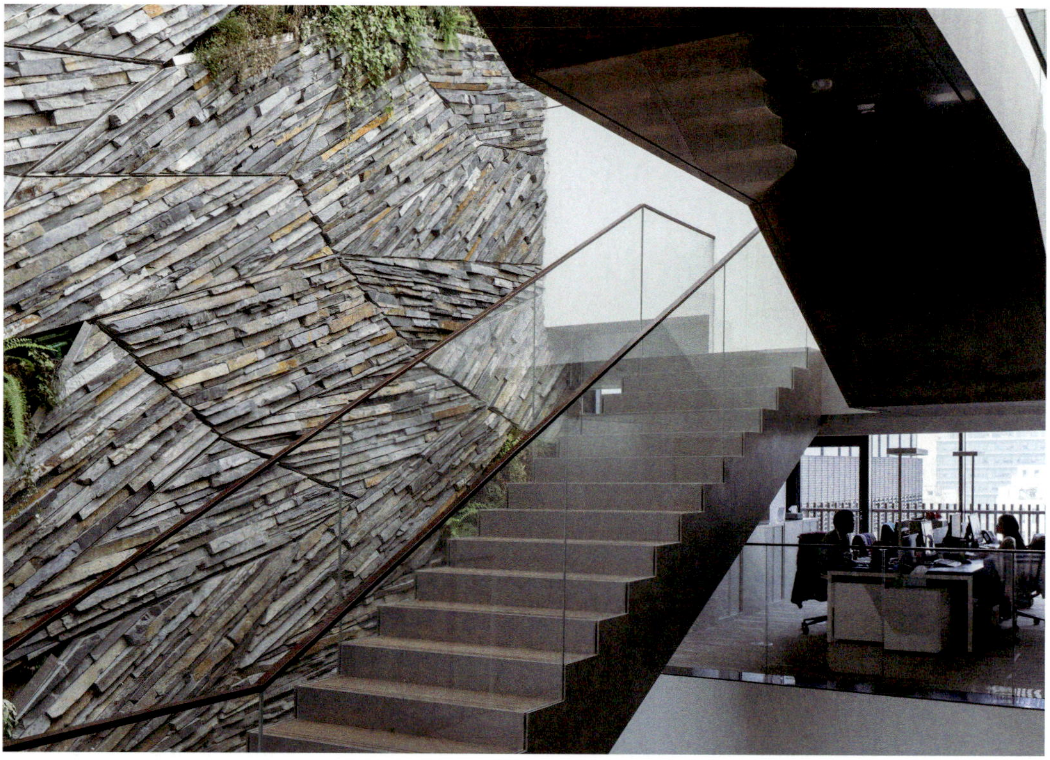

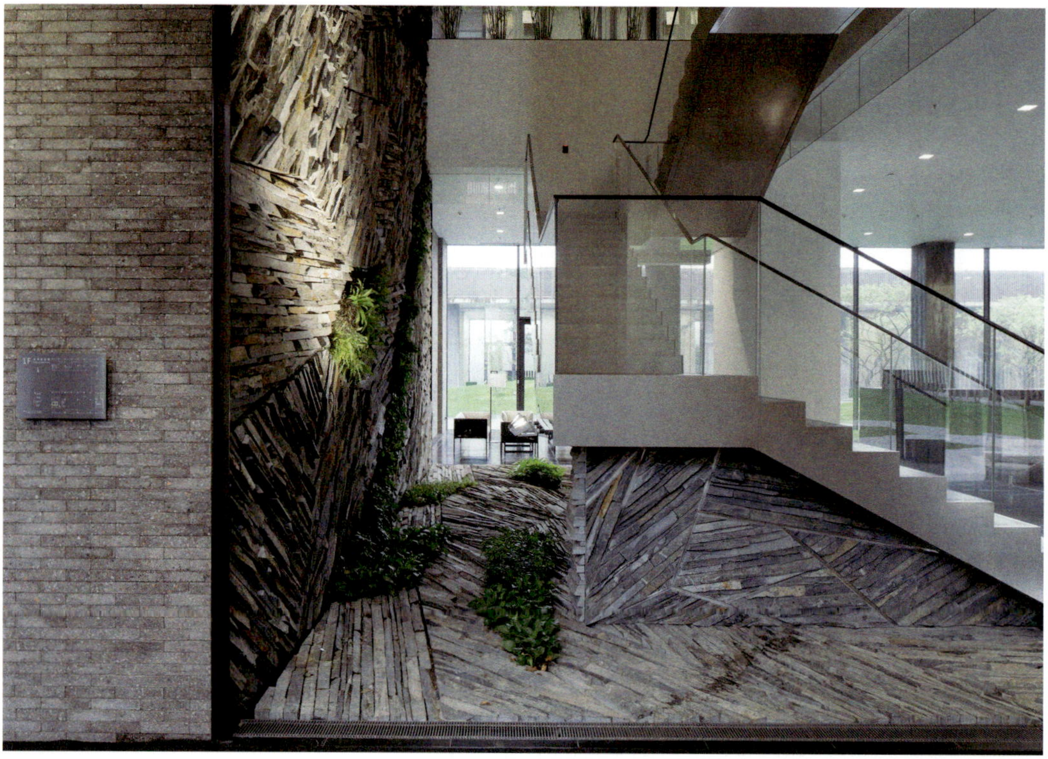

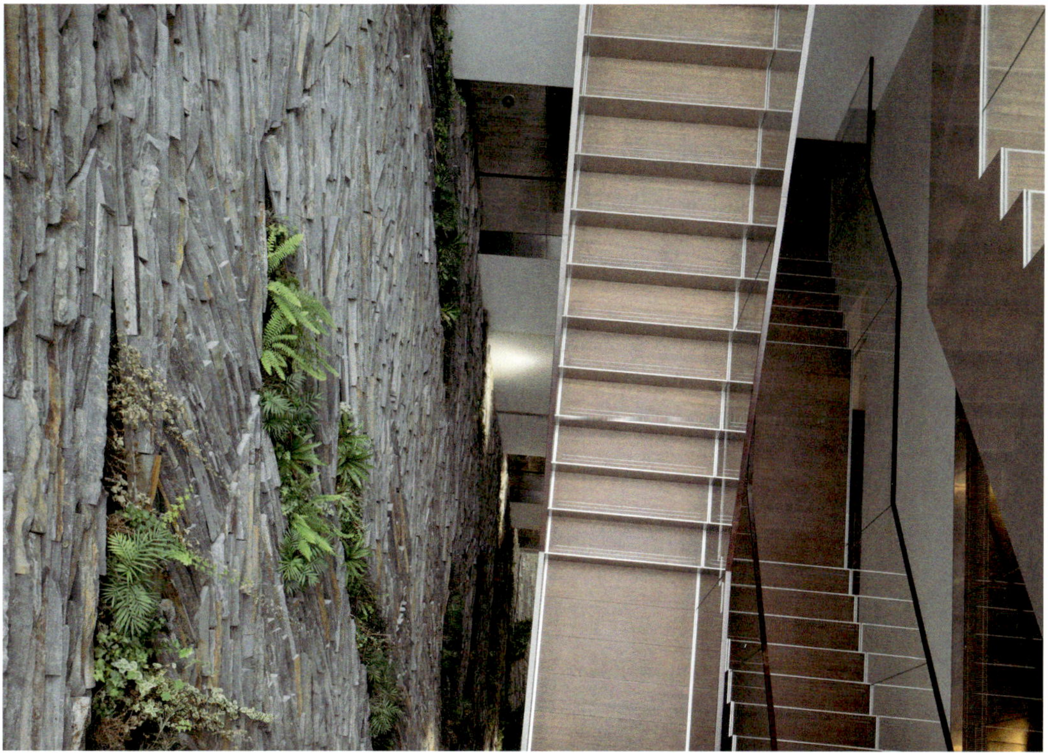

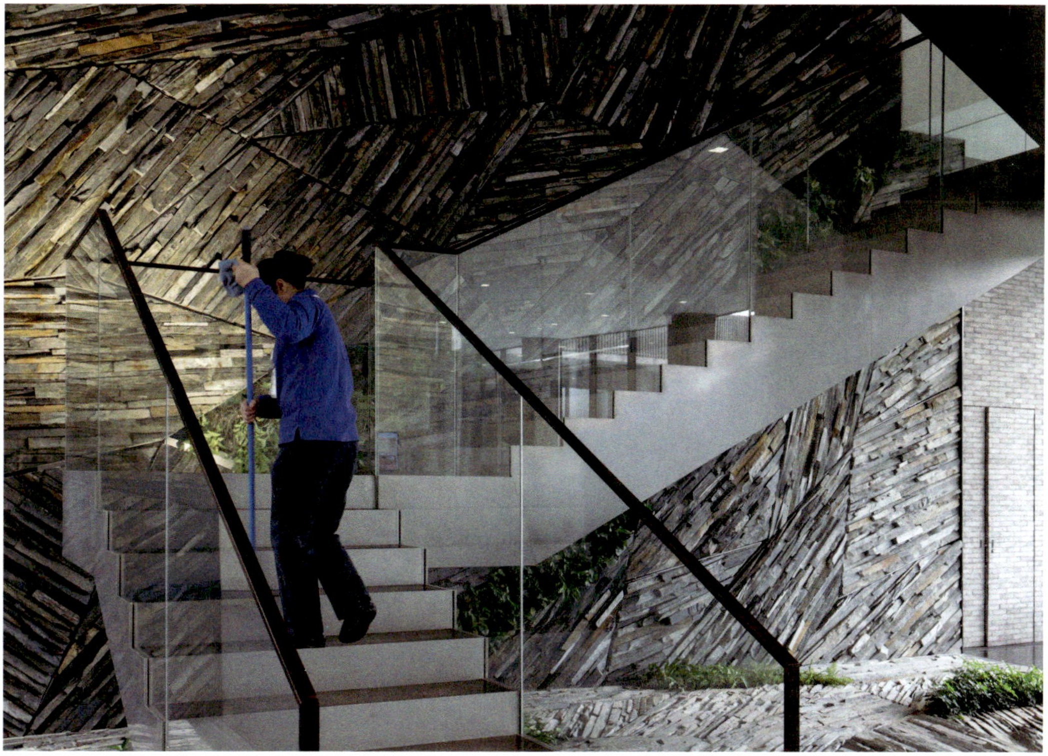

Architect	Diener & Diener Architects Basel, Switzerland	Floor area above ground	12,349 m² 132,924 sq. ft.
Workplace design	Sevil Peach	Floor levels above ground	6
		Building height above ground	35 m 114 ft., 10 in.
		Workplace capacity	200
		Materials and structure	Ornamental precast concrete elements, glass curtain wall

Roger Diener, born 1950, studied architecture at the Swiss Federal Institute of Technology in Zurich (ETHZ) and the Swiss Federal Institute of Technology Lausanne (EPFL). From 1999 to 2015 he was a professor at the ETH Zurich. In 1980 he established the office Diener & Diener Architects with his father, just four years after joining the office Marcus Diener Architect, the office his father formed in Basel in 1942.

Diener & Diener takes a keen interest in creating new types of buildings, and the office has developed various urban design plans to restructure large industrial areas. To Diener & Diener, "The House and the City" is the field of debate and exchange. When working on architectural projects in culturally sensitive contexts they define the correlation between the building program, the place, and its roots in the depths of time.

280

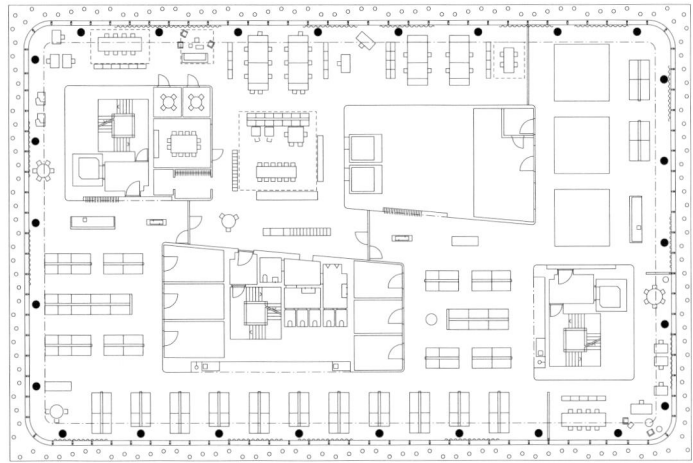

Building 3, L2

0   5   10   20 m

ROGER DIENER

BUILDING 3

The laboratory building design is influenced by Asian and Western architecture, a condition found in Shanghai since the nineteenth century. The exterior wall consists of two layers. The inner layer is a glass curtain wall and inner load-bearing structure. The outer layer is framed by "dancing columns" on an over-hang. The undulating form of the dancing columns is inspired by a preproduct of traditional Chinese roof tiles. The positioning of these dancing columns varies in depth and width. This softly and lightly defined outer layer gives the building a continuously changing appearance, depending on light and viewing position. It not only makes an architectural impression but also protects the interior from direct sunshine, while the large windows of the inner layer enable users to profit from indirect natural light and allow a generous view to the outside.

The interior is designed as an interactive, collaborative space. An open environment engenders communication, but at the same time respects the need of the users to concentrate on their work by giving them the opportunity to withdraw for undisturbed conversation. The furniture and material selection confer a warm atmosphere to the otherwise loft-like laboratory character.

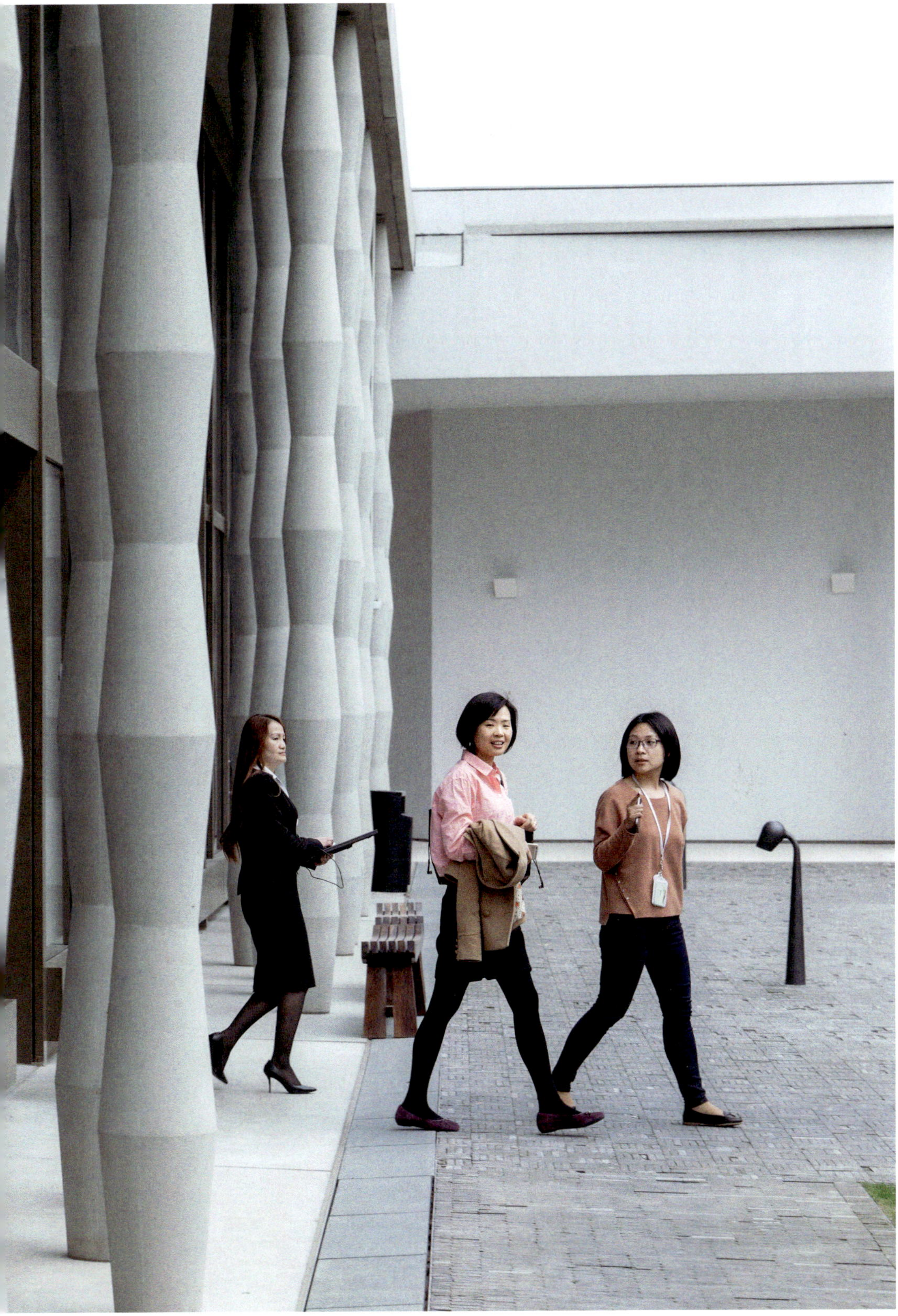

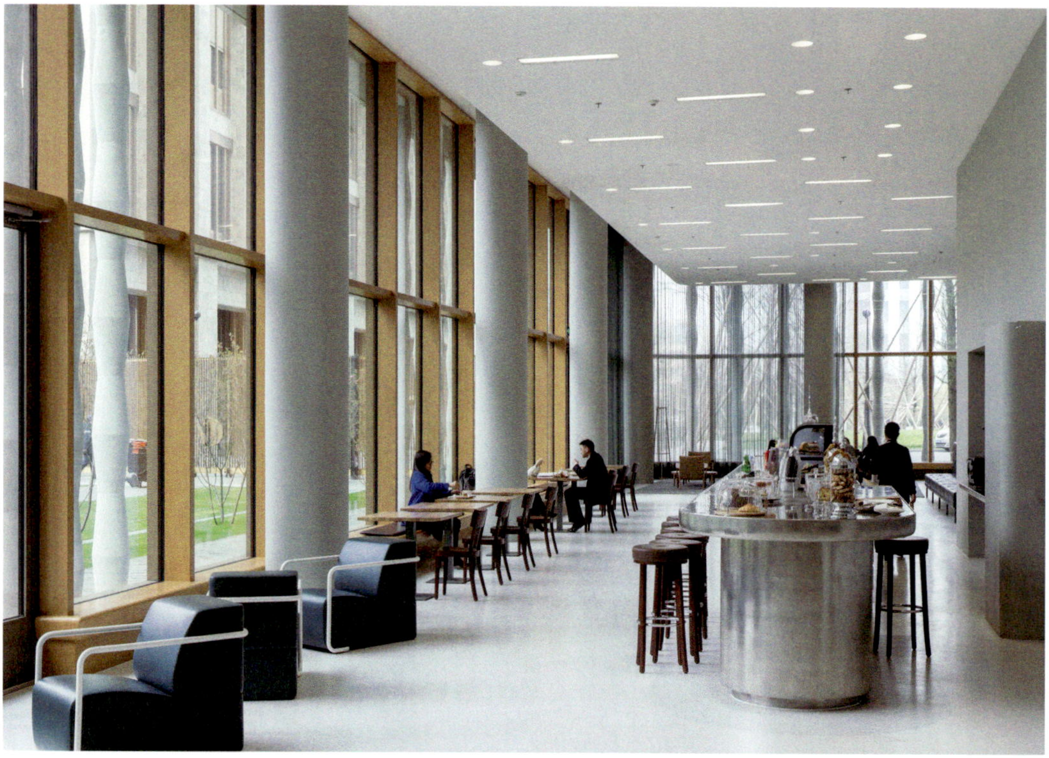

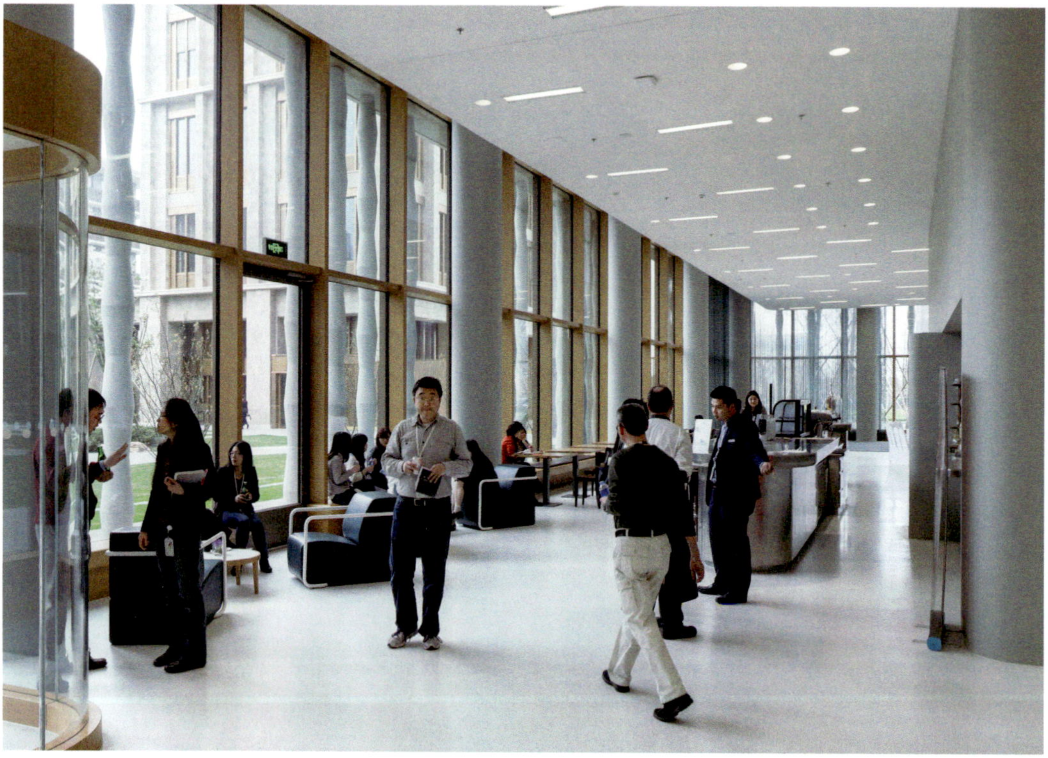

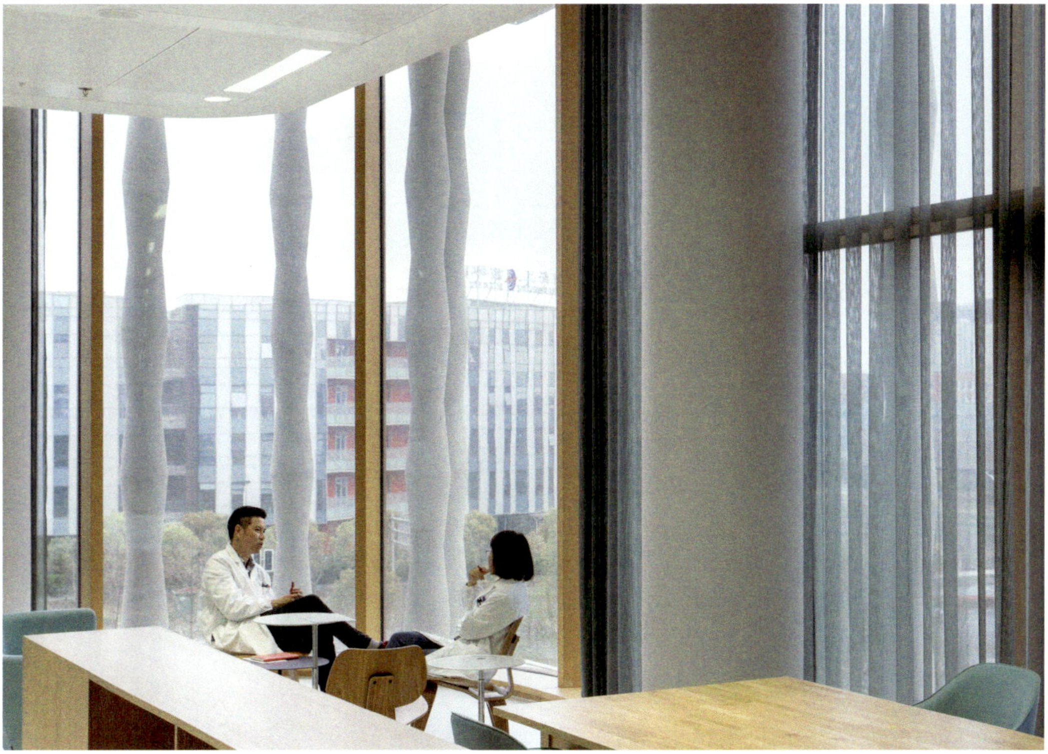

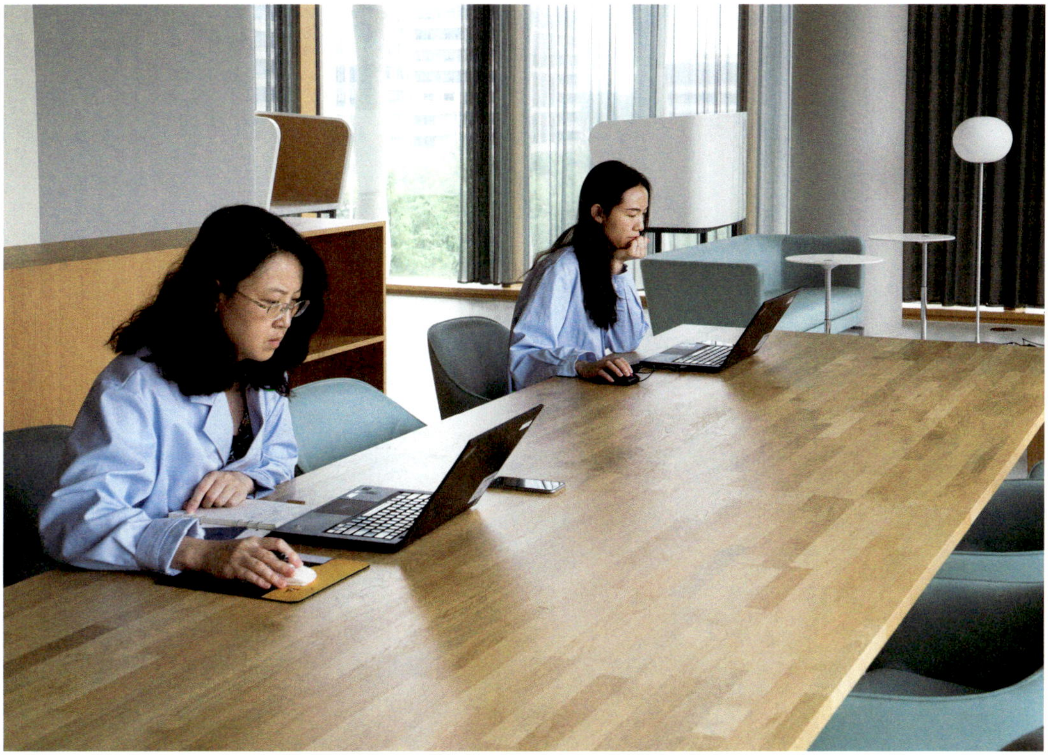

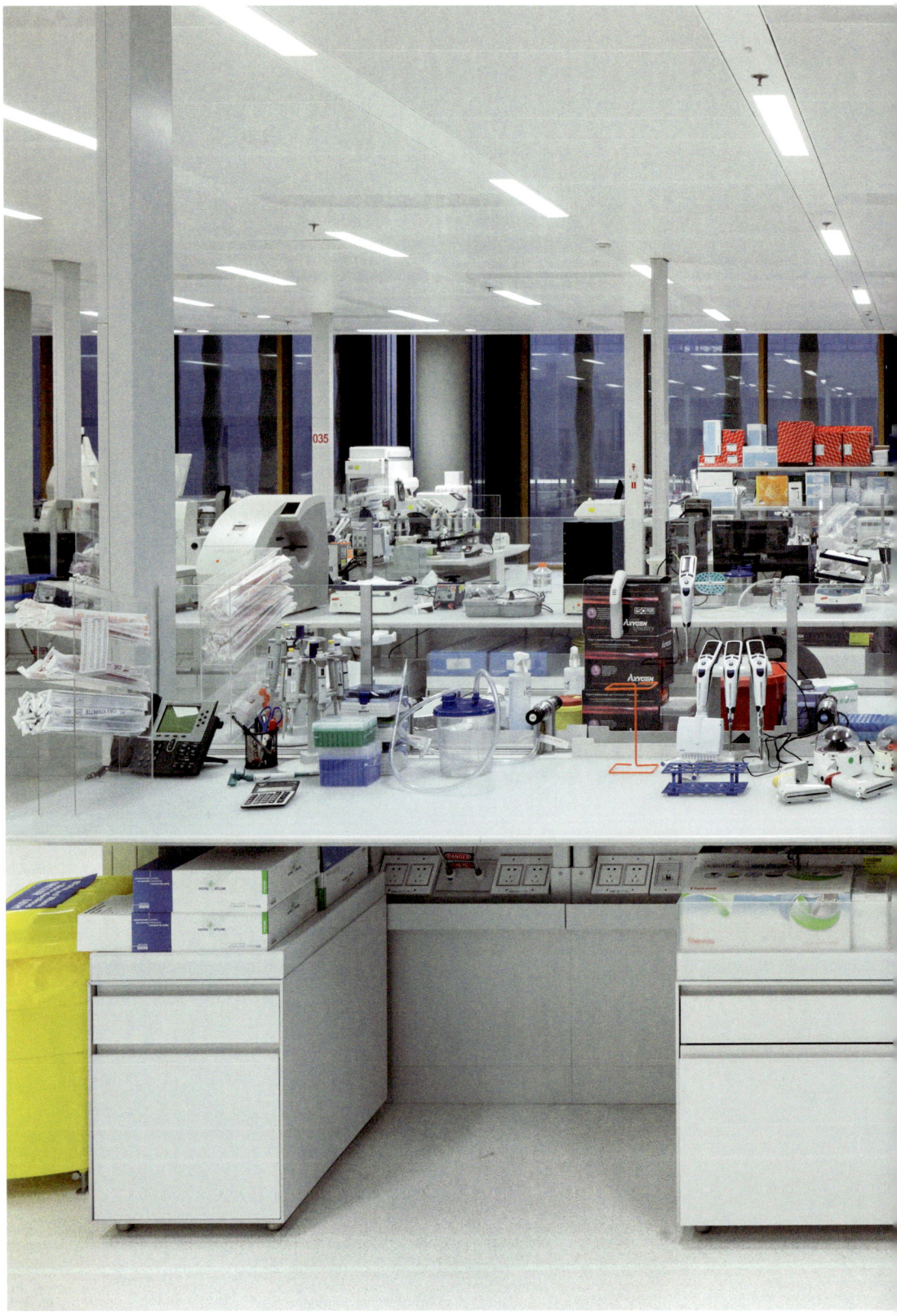

Architects	Atelier FCJZ Beijing, China	Floor area above ground	14,127 m² 152,062 sq. ft.
Project architect	Simon Lee	Floor levels above ground	6
Landscape Design	Beijing Tsinghua Tongheng Urban Planning & Design Institute	Building height above ground	36.45 m 119 ft., 7 in.
		Workplace capacity	247
		Materials and structure	Aluminium spandrel with terracotta rainscreen and sun shading, reinforced concrete structure

Yung Ho Chang, born 1956, together with Lijia Lu founded Atelier FCJZ in 1993. Both Chang and Lu are originally from Beijing and were educated in China as well as in the United States. Besides the practice, Chang has taught at various universities in the United States and in China. From 2005 to 2010 he was the head of the architecture department at MIT. Currently, he is a professor at MIT and Tongji University in Shanghai.

Atelier FCJZ sees itself as an open laboratory as much as a design practice; its range of research interests spans from material to lifestyle. Such an attitude has resulted in an operation that defies definition and is fully multidisciplinary with outputs ranging from planning, architecture, furniture, products, and stage to clothing and jewelry.

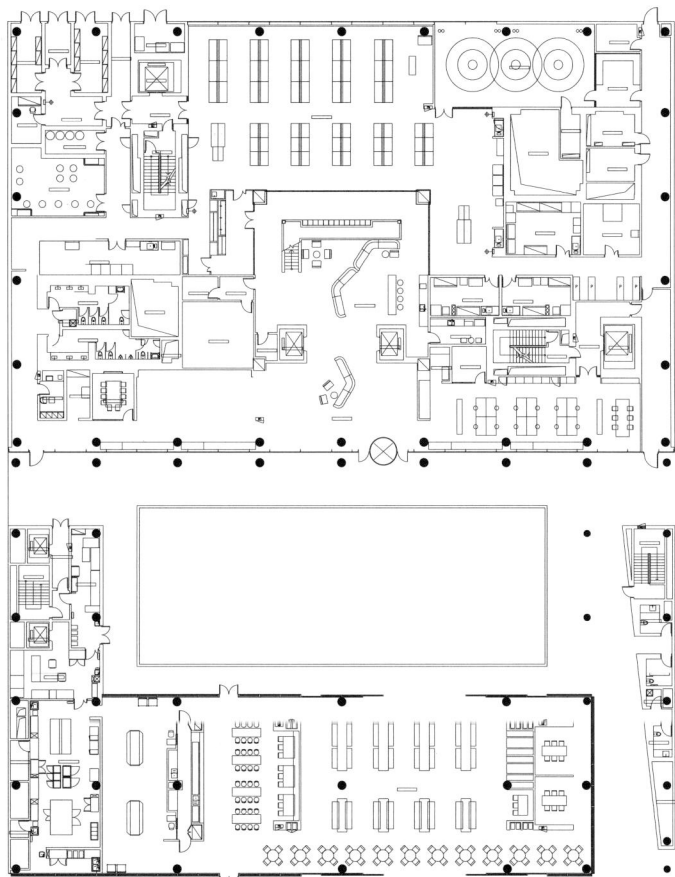

Building 4, L1

YUNG HO CHANG

BUILDING 4

The laboratory building concept is based on the characteristic gray clay architecture of the Yangtze River Delta region. In this aspect, the façade is designed as a unitary, modular terracotta system, giving the building a simple and refined but rich expression. A canteen is situated on the ground floor with access to the main landscape, as well as to the internal courtyard. Lab and nonlab work spaces are located on all five floors of the tower building.

Integrated lab and non-lab spaces surround the bamboo-clad central staircase, which is designed to foster interaction through the openness of the central void, interrupted only by "floating" platforms for escape or small meetings. Laminated bamboo is used to highlight informal meeting areas throughout the building.

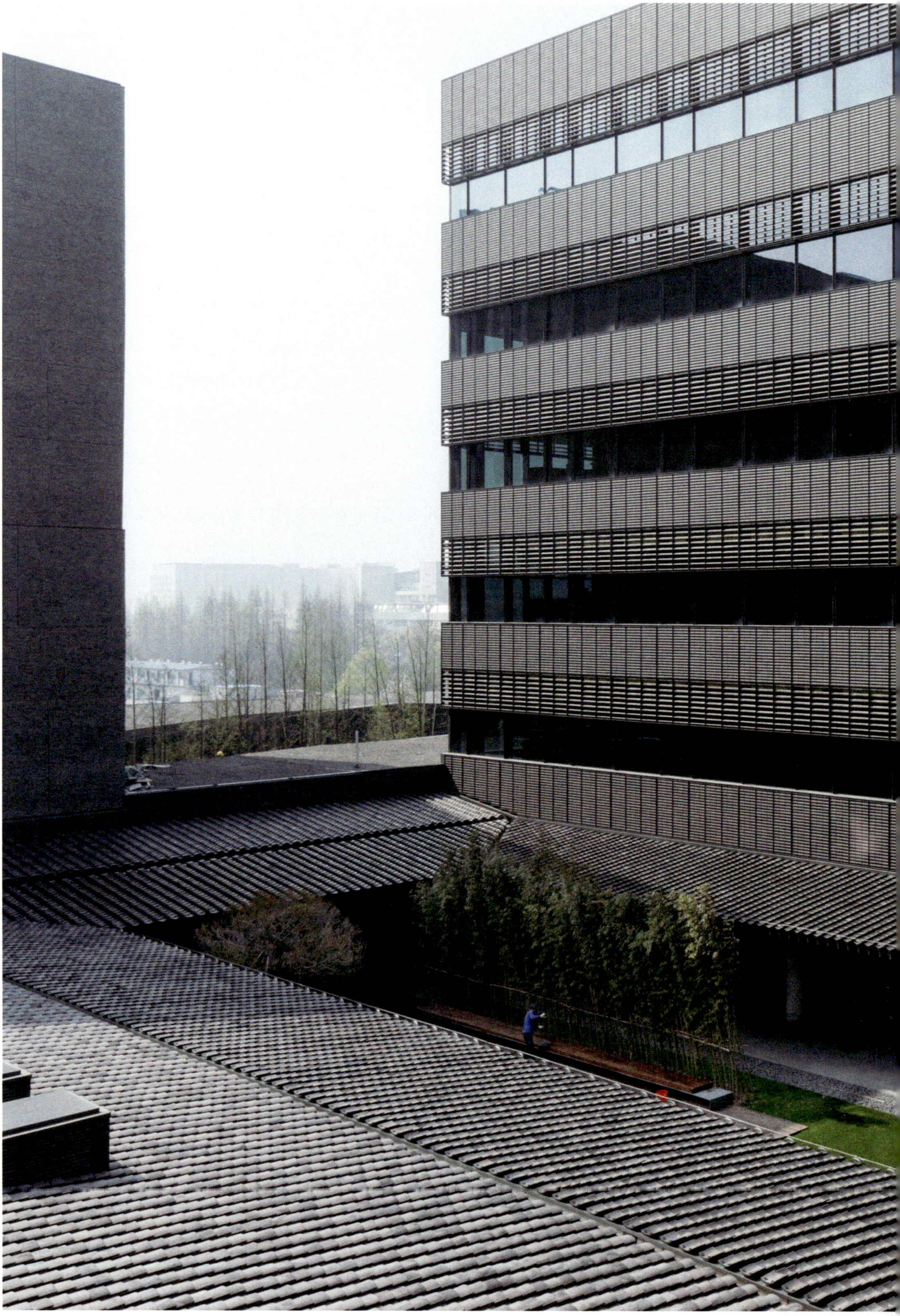

296

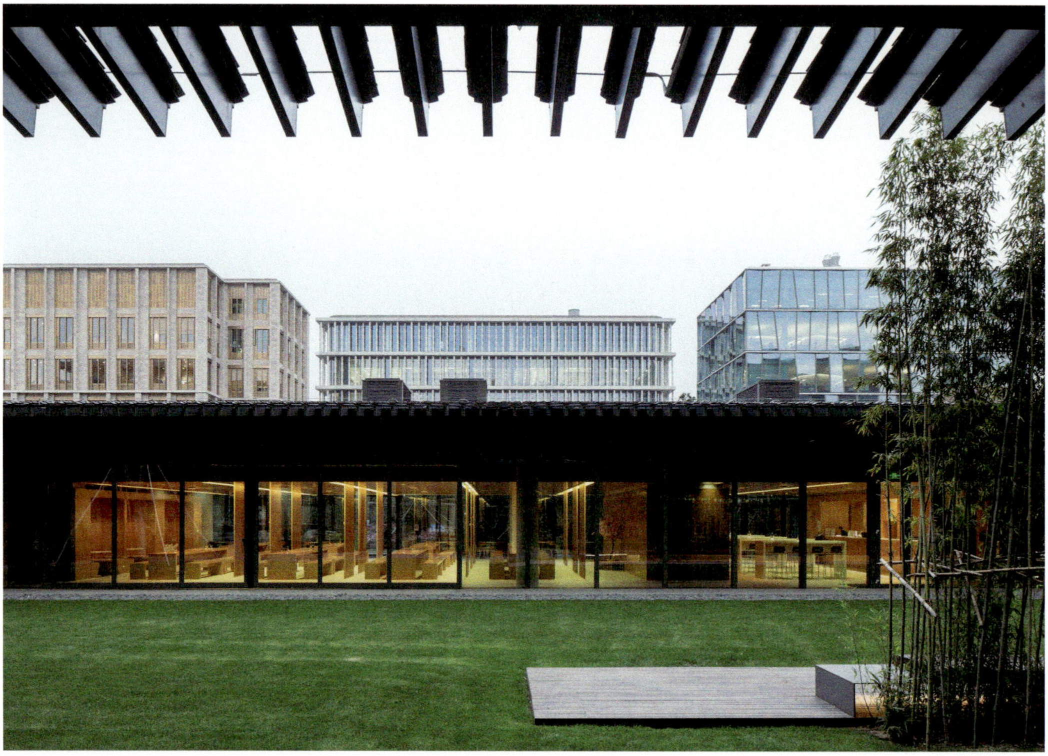

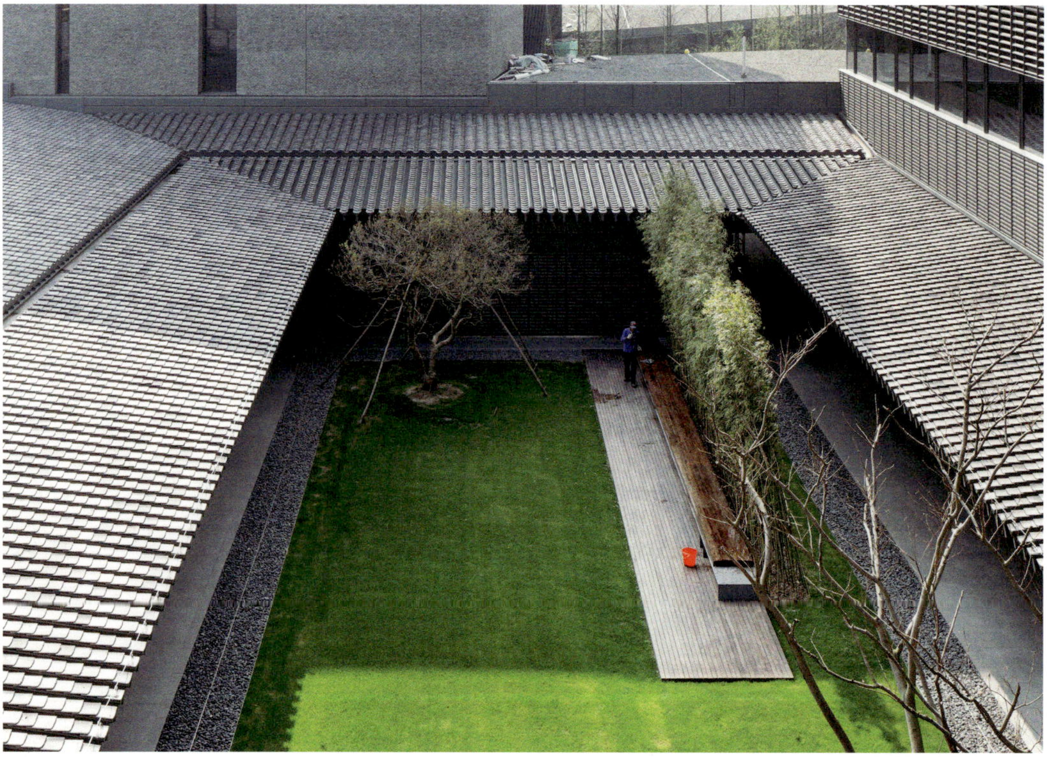

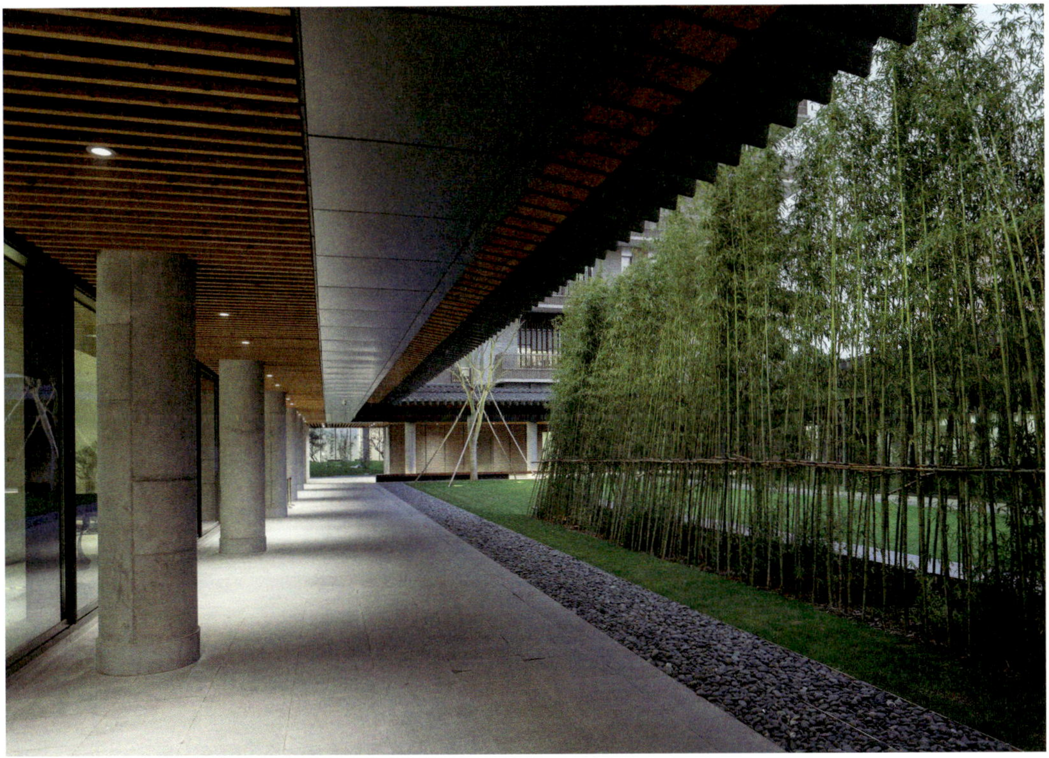

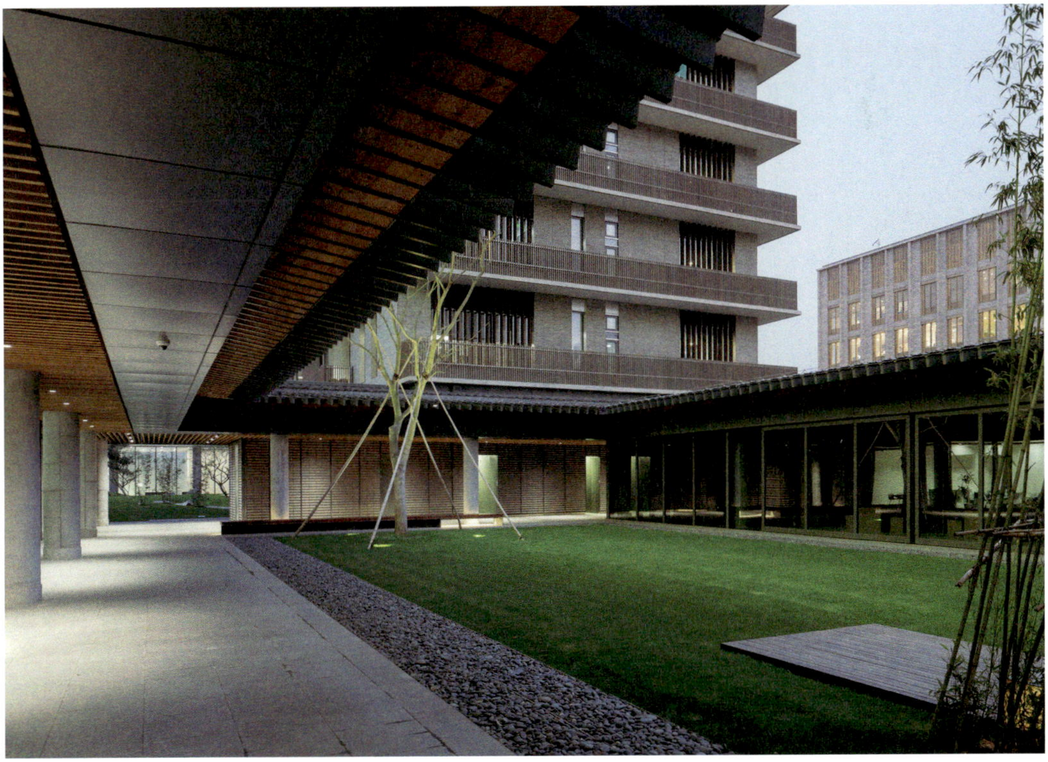

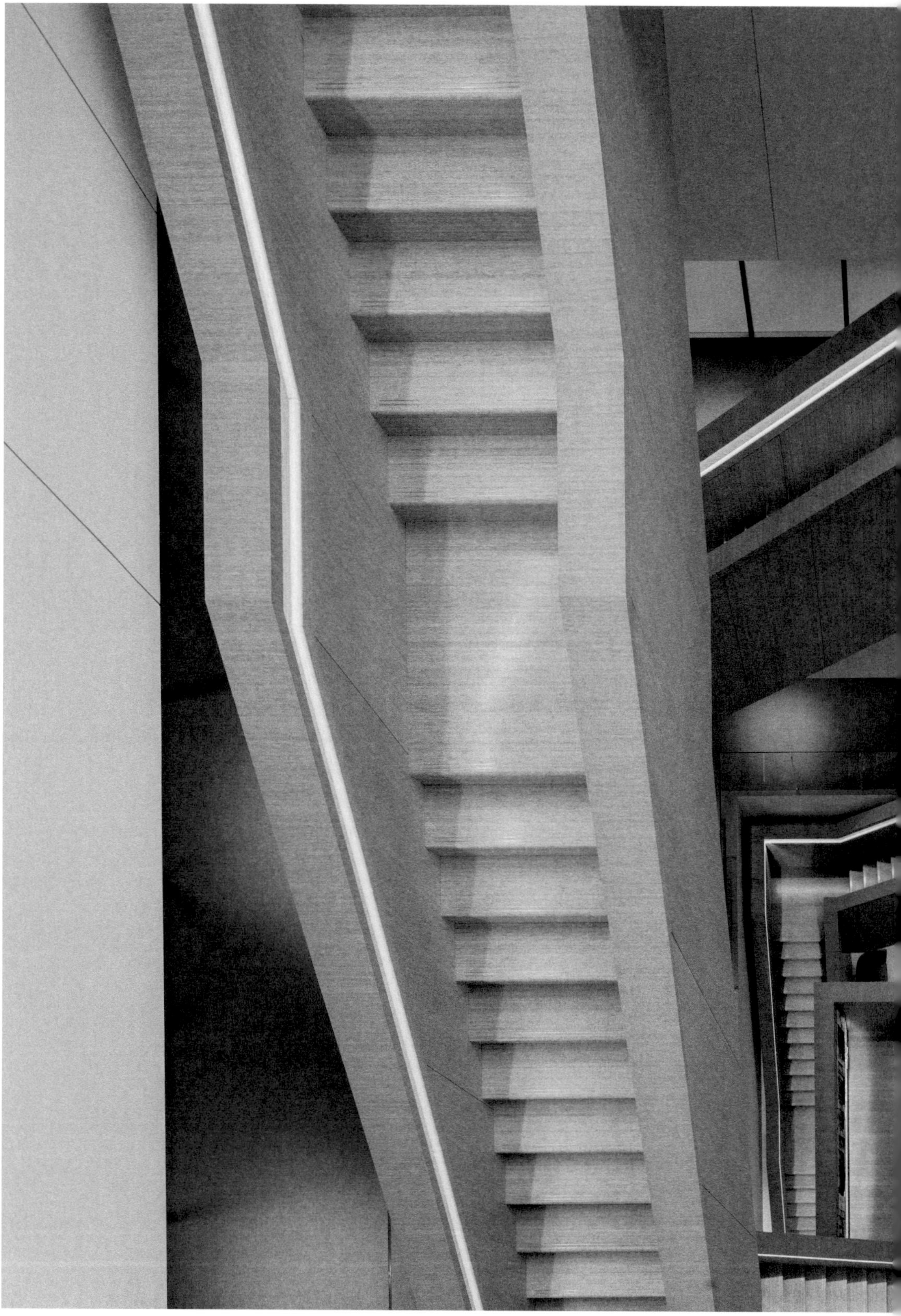

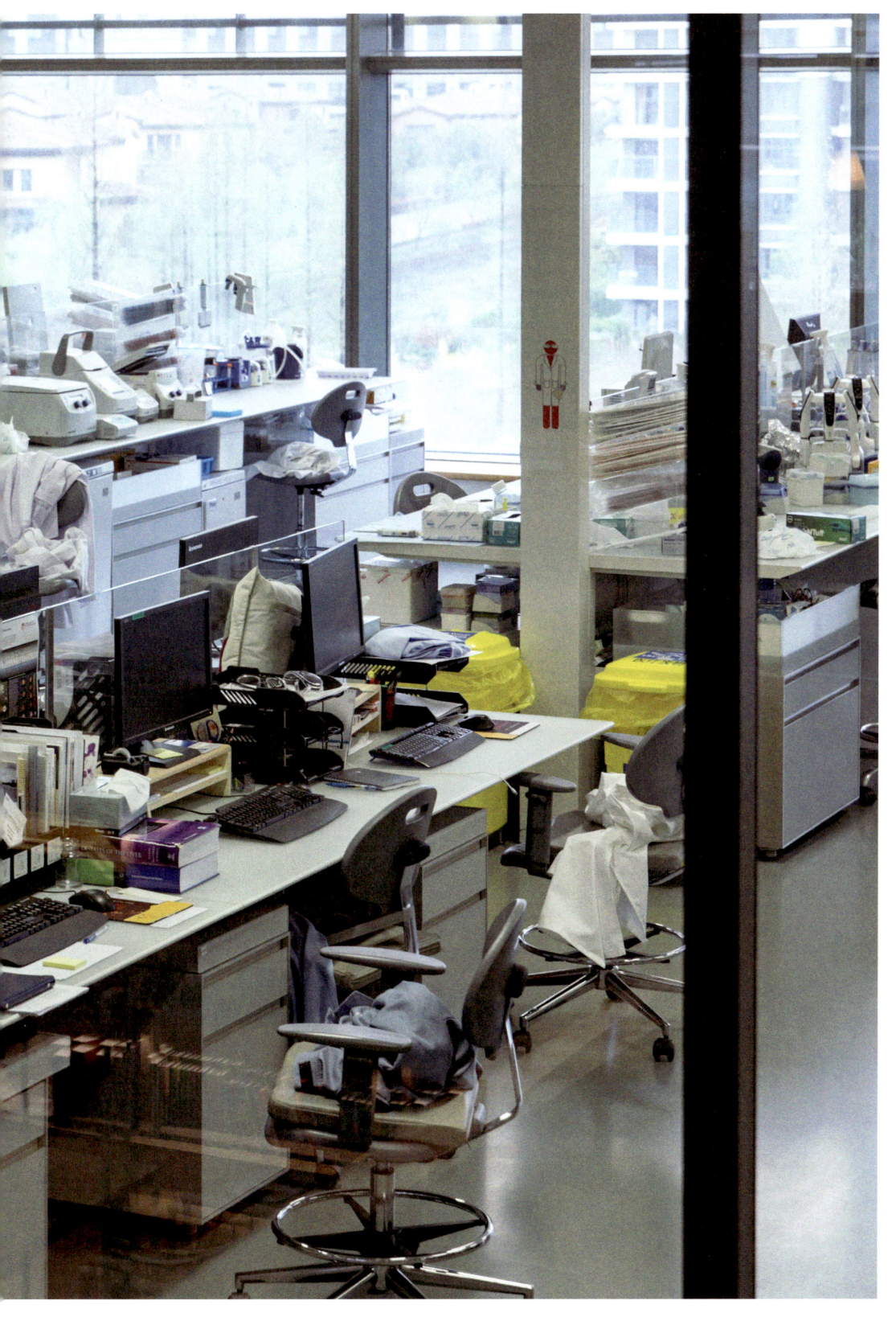

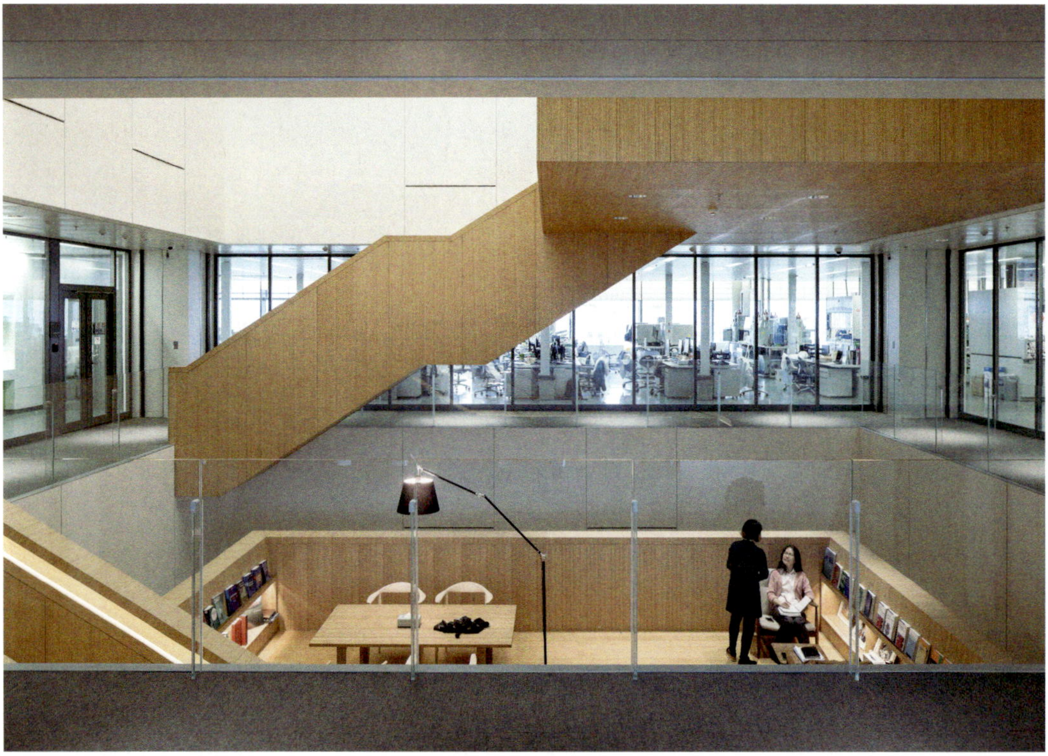

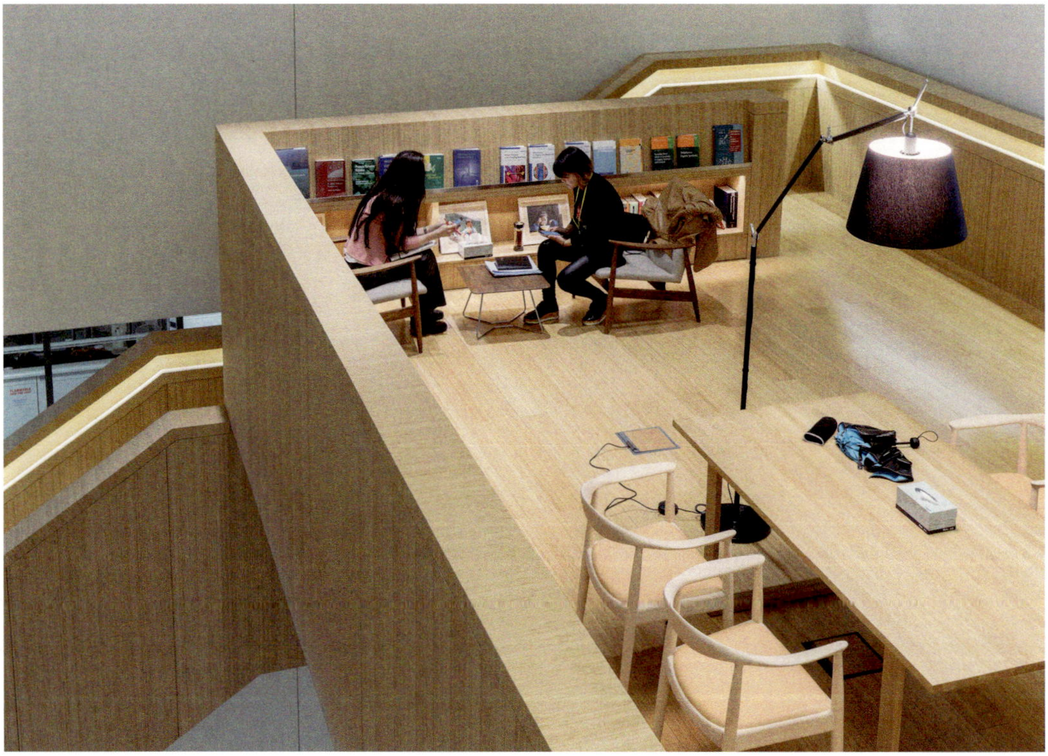

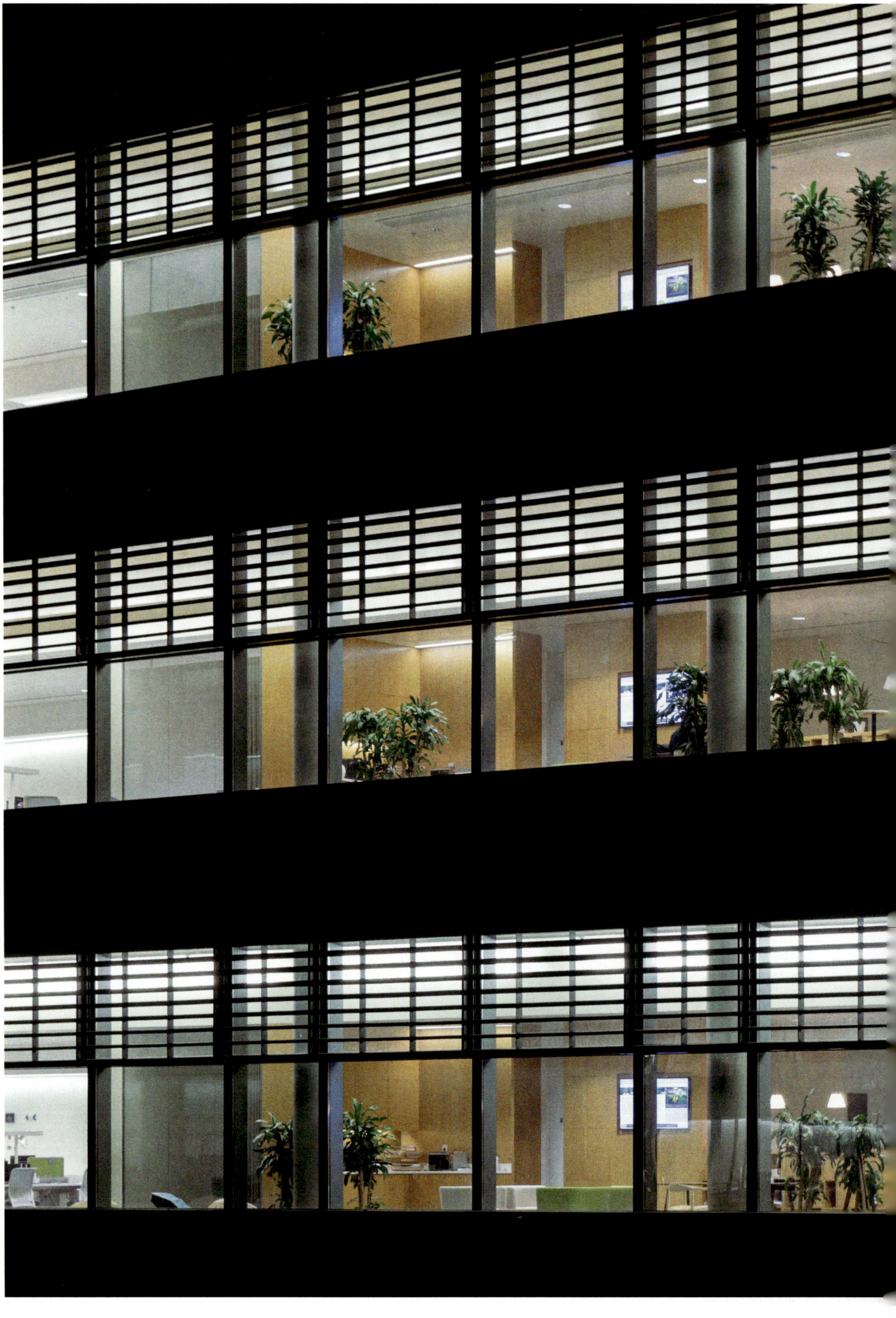

Architects	ZAO/standardarchitecture Beijing, China	Floor area above ground	8,700 m² 93,646 sq. ft.
Project architect	Zhang Ke	Floor levels above ground	6
Landscape Design	Teresa Moller	Building height above ground	32 m 105 ft.
		Workplace capacity	254
		Materials and structure	Reinforced concrete structure, glass curtain wall

Zhang Ke, born 1970, received his master's degree in architecture from Harvard University's Graduate School of Design in 1998 and his master's and bachelor's in architecture from Tsinghua University in Beijing. In 2001 he founded his studio ZAO/ standardarchitecture.

With a wide range of realized works, the studio has emerged as one of the most critical and innovative protagonists among the new generation of Chinese architects. Although Zhang Ke's completed works produce exceptionally provocative visual results, his buildings and landscapes are often rooted in their historical and cultural settings with a degree of intellectual debate.

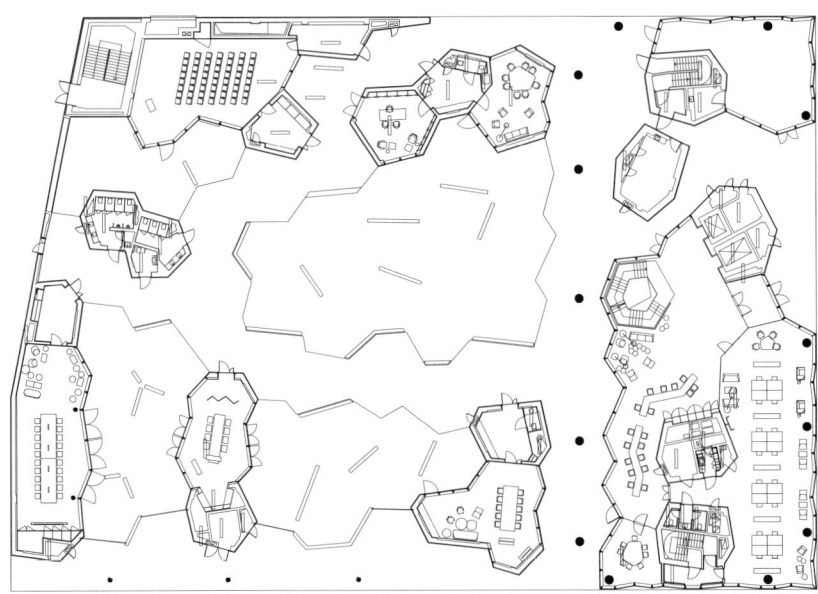

Building 5, L1

0   5   10   20 m

ZHANG KE

BUILDING 5

The office building is based on the space-forming concept of cellular structures. Three-dimensional cells comprise the outlines of the complete building volumetric, some filled and some retained as voids. This space-generating concept projects itself onto the floor plans and the building's envelope, with its tilted glass panes. In the courtyard the organic grid system generates a system of cellular pockets of indoor and outdoor spaces.

The interior prioritizes lightness as an overall value: lightness of space, lightness of structure, and lightness of mood. This concept of lightness manifests itself in the use of sparse materials and color palette throughout the overall interior design. In this simple surrounding, the furniture creates a warm and inviting environment. Each floor has selected furniture groups highlighted in a unique signal color.

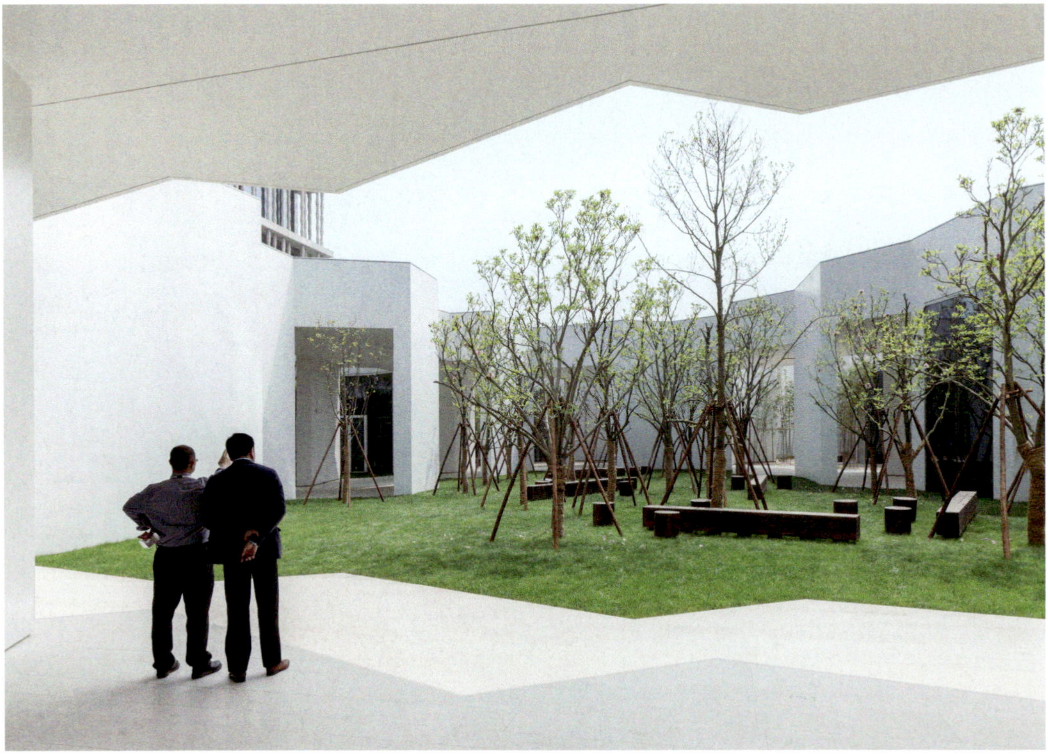

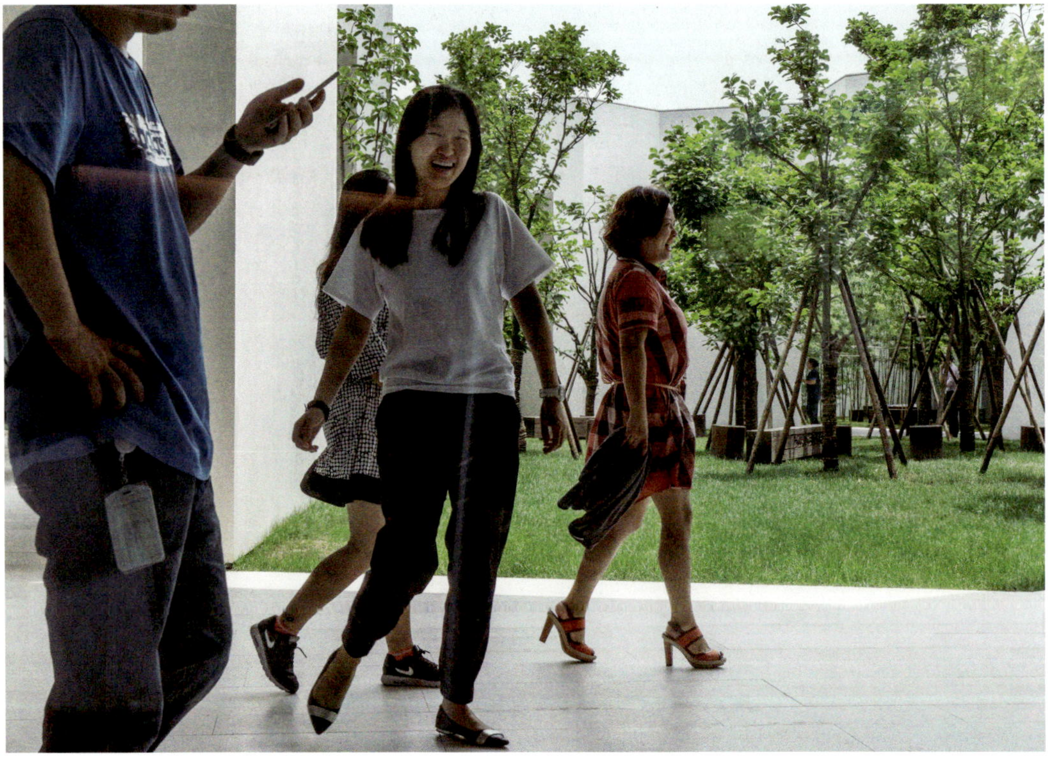

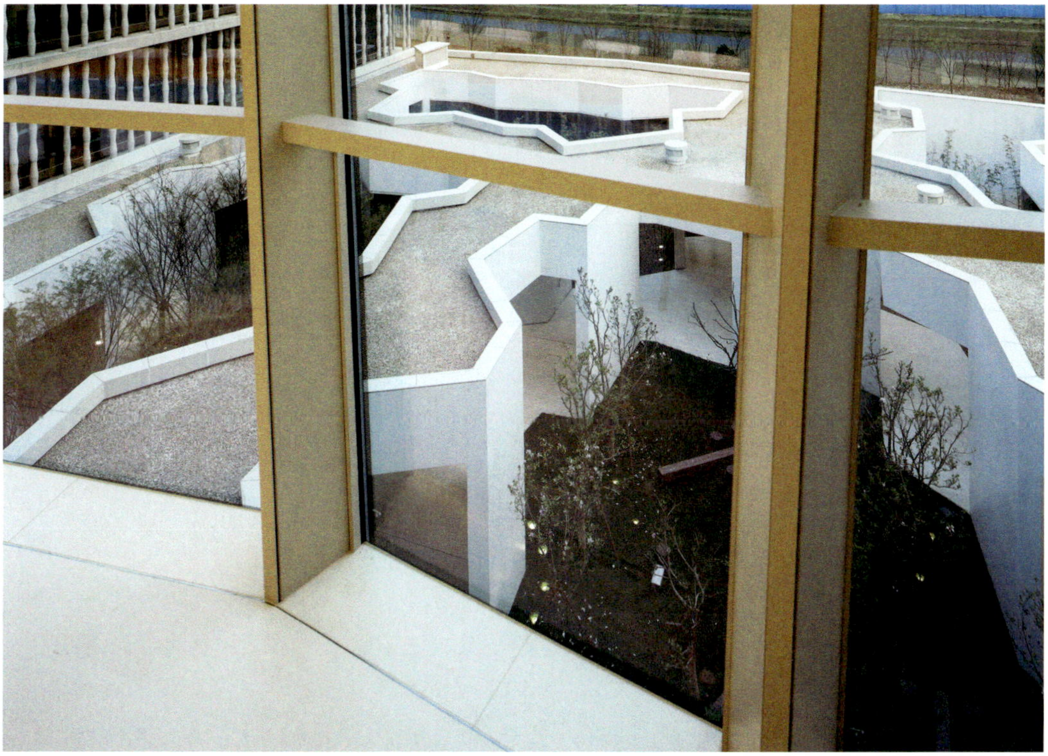

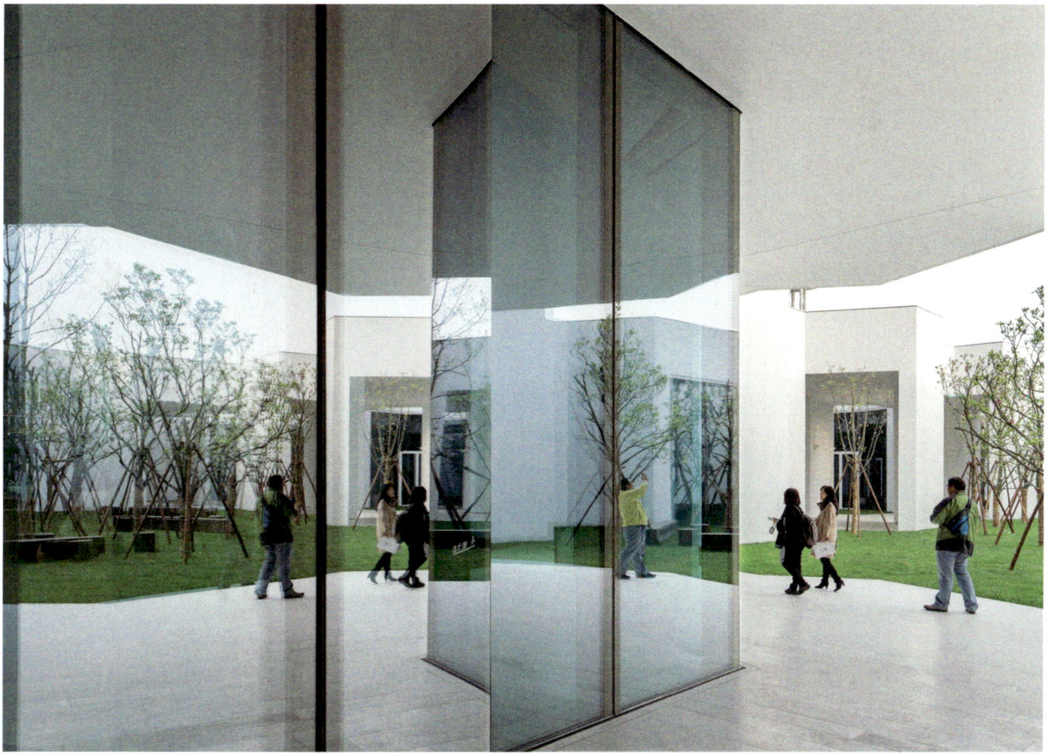

Architects	ELEMENTAL Santiago, Chile	Floor area above ground	8,322 m² 89,577 sq. ft.
Project architect	Diego Torres	Floor levels above ground	6
Landscape Design	Teresa Moller	Building height above ground	32 m 105 ft.
		Workplace capacity	204
		Materials and structure	Fair-faced concrete, reclaimed broken brick, glass curtain wall

Alejandro Aravena, born 1967, graduated from the Universidad Católica de Chile in 1992. From 2000 to 2005 he was a professor at Harvard University's Graduate School of Design, where the path to the foundation of ELEMENTAL started. He is the Pritzker Prize Laureate 2016 and was the director of the XV Venice Architecture Biennale 2016.

ELEMENTAL is a do tank founded in 2001, focusing on projects of public interest and social impact, including housing, public space, infrastructure, and transportation. A hallmark of the firm is its participatory design process, in which architects work closely with the public and final users, fostering a user-centric approach. In the development of complex initiatives, which require the coordination of multiple stakeholders, ELEMENTAL contributes innovation and quality of design.

324

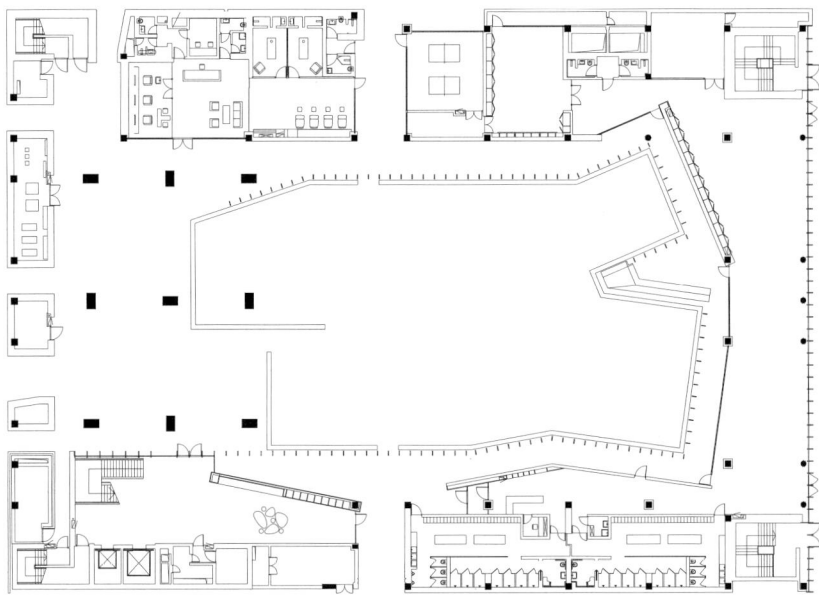

Building 6, L1

0   5   10   20 m

ALEJANDRO ARAVENA

BUILDING 6

The office building combines aspects of the local environmental and climatic conditions with the characteristic needs of an office. The relatively hermetic volume, with very controlled perforations toward the outside, addresses energy considerations as well as natural daylight requirements. Local rough, broken bricks give this monolithic building a unique texture on the exterior and internal façades.

The interior creates different spaces within the open office plan to accommodate different modes of work. Those areas are further defined by the use of varying materials and furniture. Collective work areas alongside the south façade, as well as the customized open workspaces, also offer a series of individual retreats and meeting spaces to foster interaction between users, a key to knowledge creation.

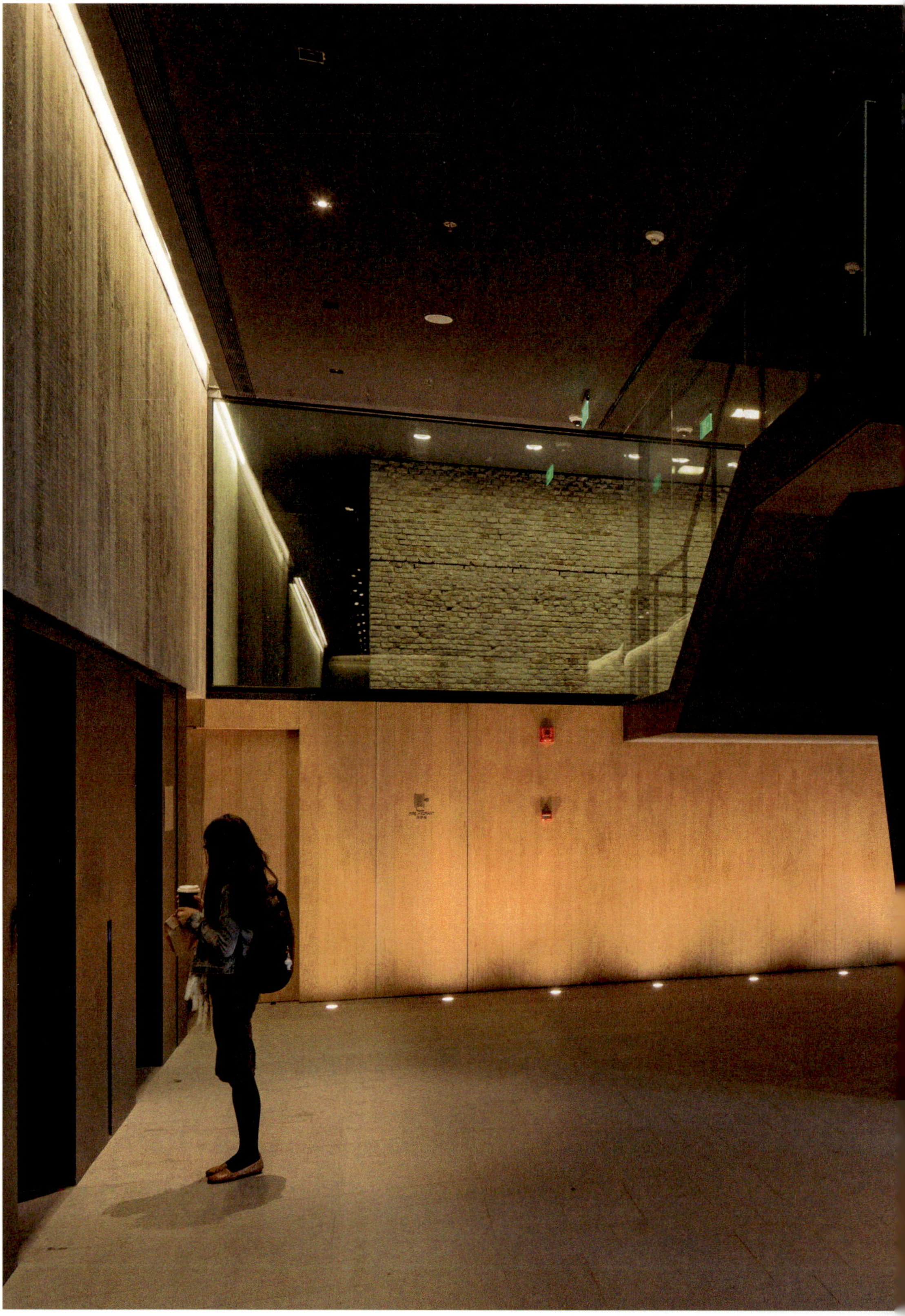

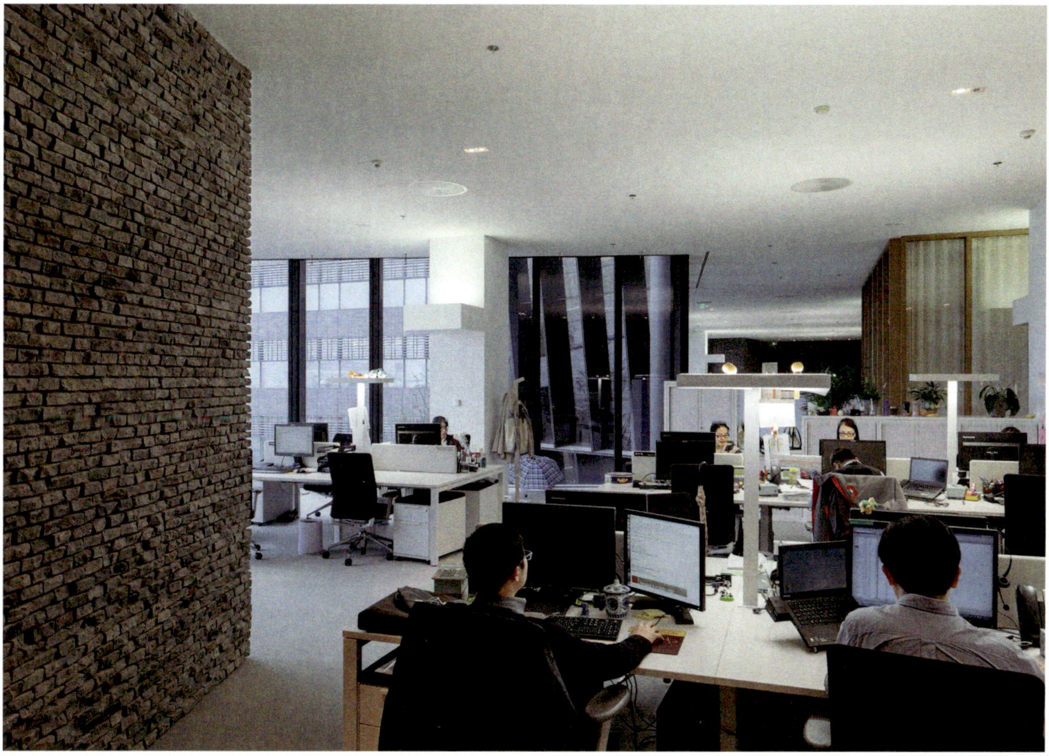

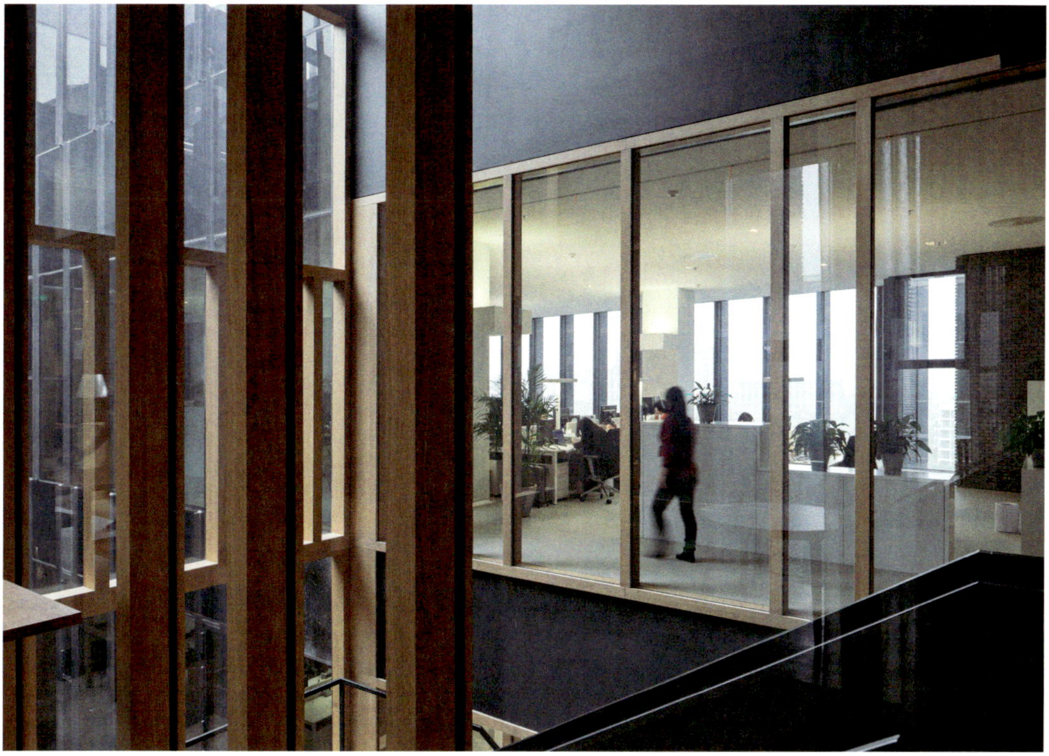

Architects	Kengo Kuma and Associates Tokyo, Japan	**Floor area above ground**	960 m² 10,333 sq. ft.
Project architect	Kengo Kuma, Hirokatsu Asano, Yuki Ikeguchi, Shengze Chen, Yencheng Chen, Tomoyuki Hasegawa, Ryota Torao, Katinka Temme	**Floor levels above ground**	2
		Building height above ground	11 m 36 ft., 1 in.
		Workplace capacity	220
		Materials and structure	Glue-laminated wooden structure, extensive greenroof, glass curtain wall

Kengo Kuma, born 1954, established Kengo Kuma & Associates in 1990. Since 2009, he is a professor at the Graduate School of Architecture, University of Tokyo, and is a visiting professor at the School of Architecture, University of Illinois at Urbana-Champaign (Chicago, USA). Kuma is also a prolific writer; many of his recent works have been published in English.

In many of Kengo Kuma's projects, attention is focused on the connection of spaces, on the segments between inside and outside, and one room to the next. The choice of materials stems not so much from an intention to guide the design of the forms but rather to conform to the surroundings, from a desire to compare materials yet show the technical advances that have made new uses possible.

336

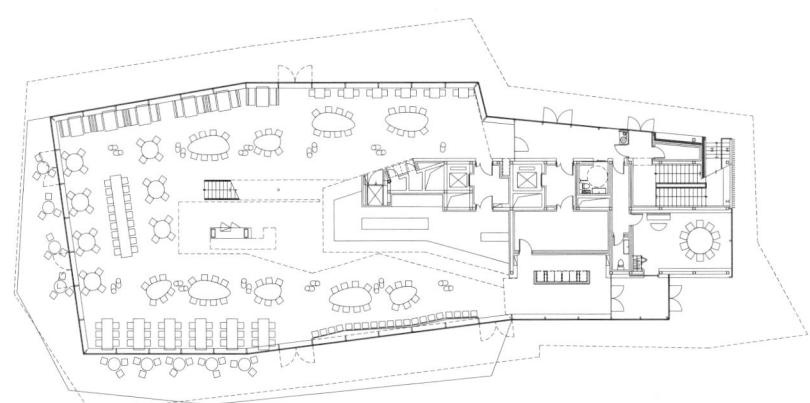

Building 7, L1

0   5   10   15 m

KENGO KUMA

BUILDING 7

The multifunction building is located in the central courtyard of the campus to provide space for dining and for meetings.

In contrast to the large-scale volumes of other buildings on the campus, this building is designed to feel like a small "house" in the middle of the campus for people to meet and gather. It is the only low-rise building in the central courtyard, and the *origami*-inspired green roof, visible from all the surrounding buildings, continues and emphasizes the theme of botanical life and sustainability that is integrated into the garden of the campus.

337

The unique roof geometry is made using the technique of glue-laminated timber, creating a tree-like wooden structure under the roof of this "house." Sets of furniture made of different organic materials placed throughout the building encourage meetings of different styles, formalities, and size.

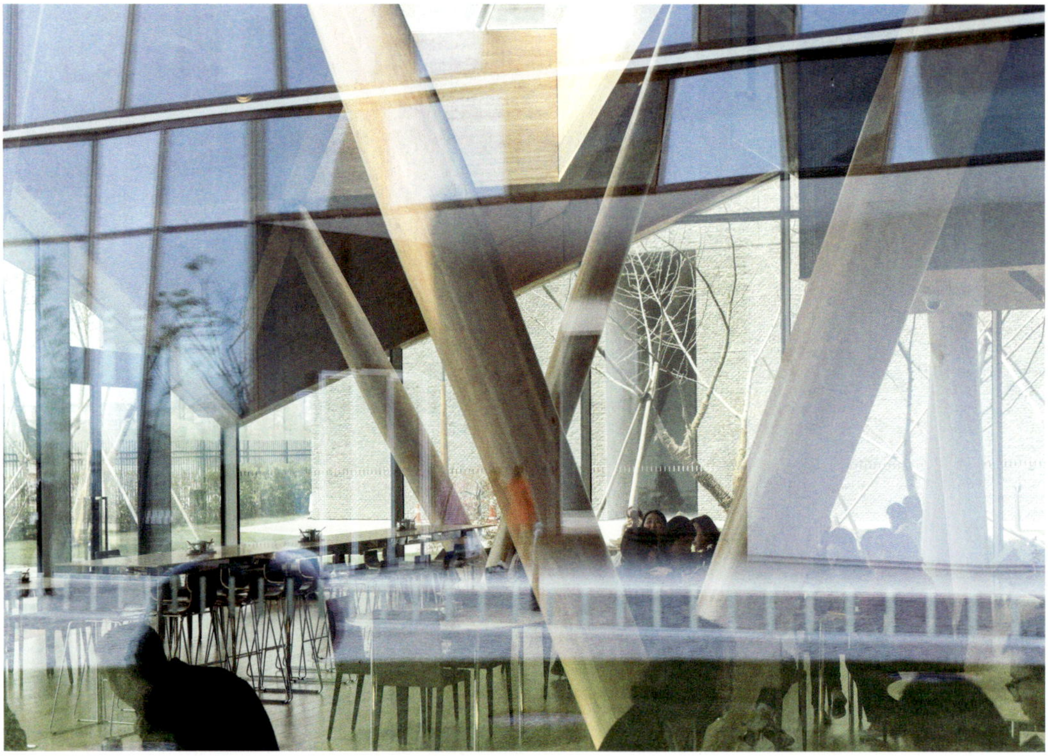

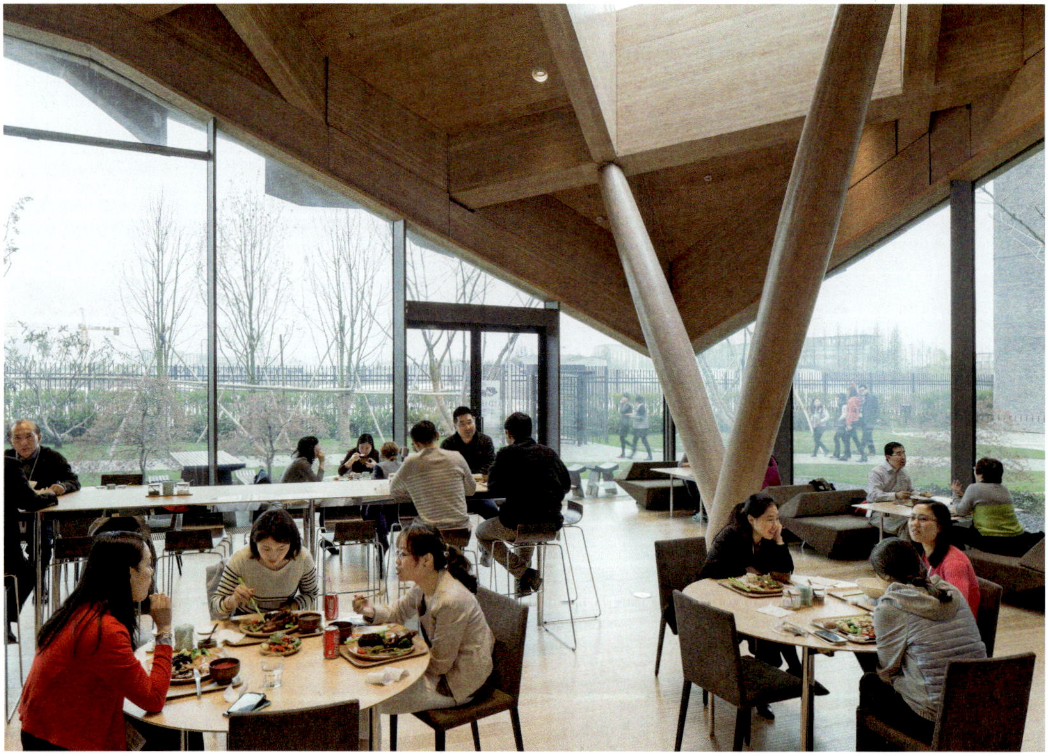

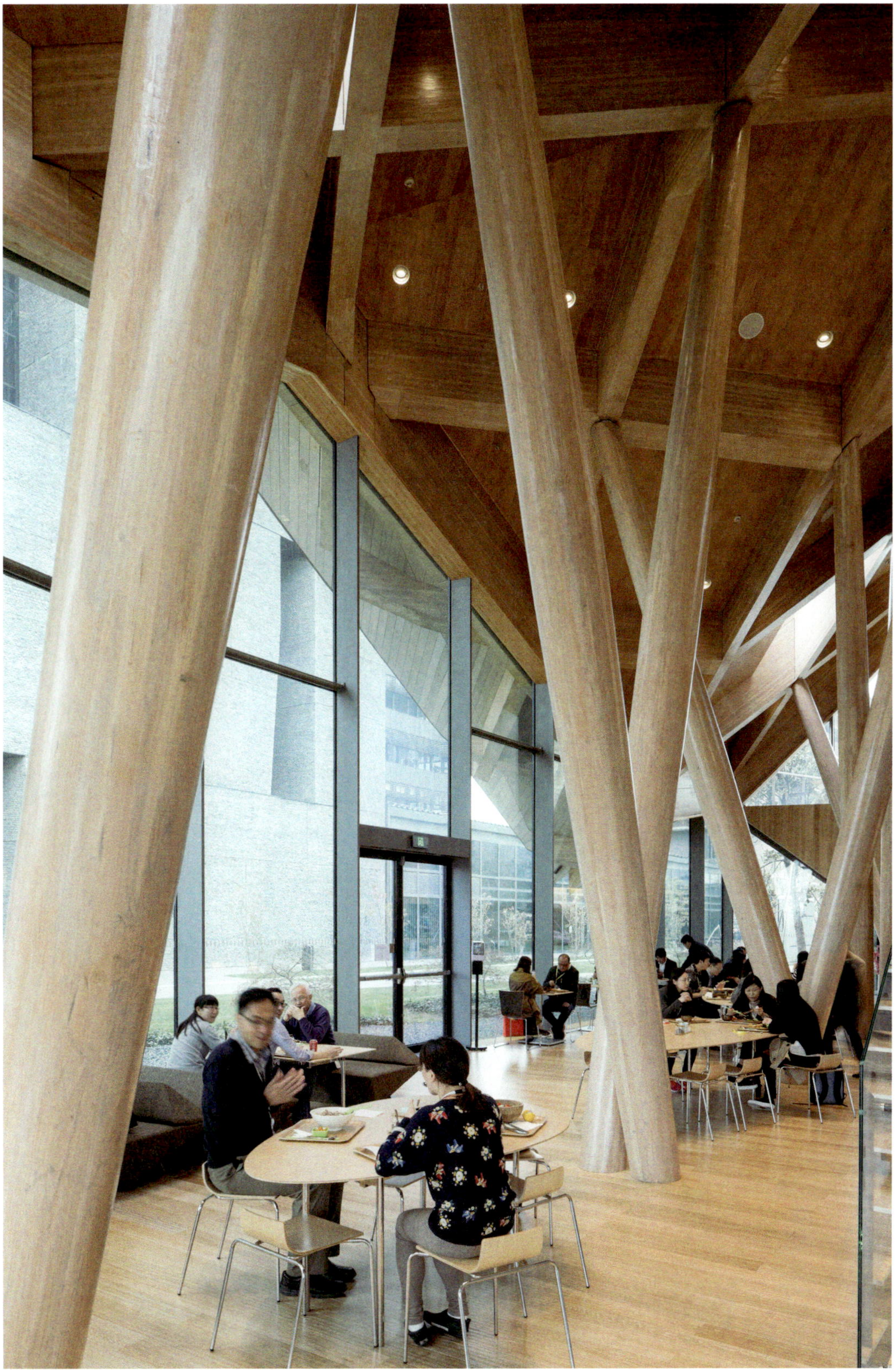

Architect	Archi-Union Architects Shanghai, China
Project Architect	Alex L. Han

Floor area under ground	101,000 m² 1,087,155 sq. ft.
Floor levels under ground	1 level + mezzanine
Basement height under ground	8 m 26 ft., 3 in.
Parking capacity (total)	950 cars
Materials and structure	Reinforced concrete structure

Philip F. Yuan founded Archi-Union Architects in 2003. The practice strives for excellence and innovation, exploring the social production and approaches to the practice of architecture in the context of modern Chinese cities; challenges traditional concepts in construction through digital design and building theories; and offers systemized solutions to customized mass production through digitally designed factory-of-the-future buildings with robotics at their core. Through years of practice, Archi-Union Architects has gradually defined its own architectural craftsmanship, means of construction, and practical philosophy of contemporary architecture. Its projects are both experimental and constructive, complete and integrated, organized with respect to design and rational with respect to construction.

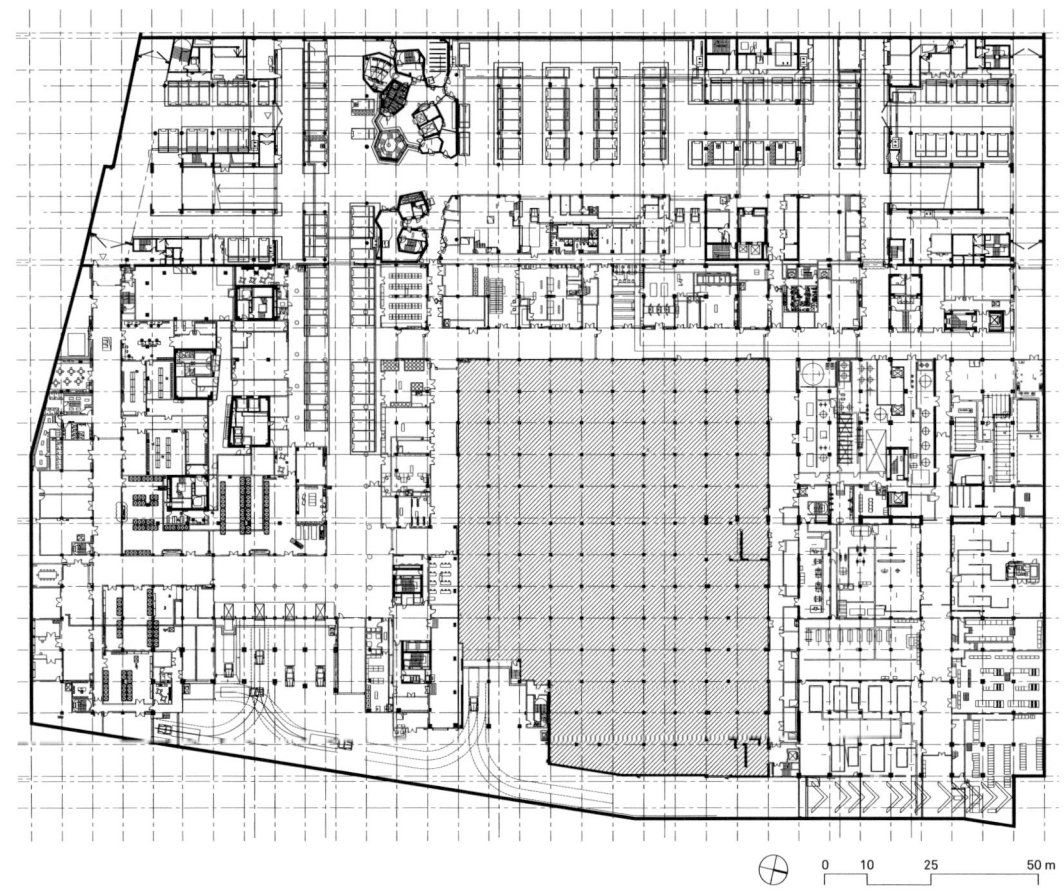

0   10   25   50 m

PHILIP YUAN

BASEMENT

The basement serves as the second entrance, offering
distinctive paths into the buildings while providing
a spatial ambiance of its own. The design utilizes the
craftsmanship of concrete combined with an abstract
rendering of traditional Chinese landscape paintings
from the Ming dynasty. This is reflected in an apparent
unbounded extension of the wall, engendering a sense
of ambiguity to the boundaries. The basement extends
throughout the master plan footprint area of the entire
site. All vehicle traffic for the entire campus enters
through the basement. In addition to parking, the basement provides centralized support functions for the
buildings, including laboratory support, and a bomb
shelter.

| **Architect** | West 8 urban design & landscape architecture Rotterdam, the Netherlands | **Materials and structure** | Traditional Chinese blue brick paving; hard wood shading structure and blue diabase cladding of walls and furniture elements |
| **Project architect** | Christian Dobrick | | |

Adriaan Geuze, born 1960, received a master's degree in landscape architecture at the Agricultural University of Wageningen. Internationally respected as a professor in architecture and urban design, he frequently lectures and teaches at universities worldwide. In 1987 Geuze cofounded West 8 as an international office for urban design and landscape architecture, with offices in Rotterdam, New York, and Belgium.

Over the last three decades, West 8 has established itself as a leading practice with an international team of architects, urban designers, landscape architects, and industrial engineers. With a multi-disciplinary approach to complex design issues, West 8 has extensive experience in large-scale urban master planning and design, landscape interventions, waterfront projects, parks, squares, and gardens, and it also develops concepts and visions for large-scale planning issues that address global warming, urbanization, and infrastructure.

ADRIAAN GEUZE

LANDSCAPE

The poetic inspiration for the Novartis Campus lies
in the beauty of the rural orchard, beholding a diverse
array of warm colors, gently undulating topography,
seasonality, textures, elegant detailing, and fragrant blos-
soms. A careful procession of naturally choreographed
spaces, the campus landscape gradually spirals inward
from a dramatic Taxodium entrance area through
grand bronze entry gates, along a beautifully paved
network of pathways and into a lush orchard commons
that features an iconic pavilion structure complete
with custom-made furniture. The union of these places
and the flow between them reflects the landscape's
overarching purpose of weaving together the campus'
seminal architecture.

PHOTOS BY LOCATION

IMAGE CREDITS

24 Reproduced by kind permission of the Syndics of Cambridge University Library, DAR121
26 top; 27 top © Historic England Photo Library
26 bottom The Natural History Museum / Alamy Stock Photo
27 bottom Reproduced by kind permission of the Syndics of Cambridge University Library, DAR233.2;18
29 From D'Arcy Thompson, *On the Growth and Form,* 1961/2010, p. 109 © Cambridge University Press, reproduced with permission
30 left © Curt-Engelhorn-Stiftung / Helmut-Gernsheim-Archiv
30 right From D'Arcy Thompson, *On the Growth and Form,* 1961/2010, p. 173 © Cambridge University Press, reproduced with permission
40 Bettmann / Contributor
43 Francis Crick Papers, PP/CRI/H/1/16, Wellcome Library, London
44 © Christiane Nüsslein-Volhard
51 From Prof. Dr. Otto Wilhelm Thomé, *Flora von Deutschland, Österreich und der Schweiz*, 1885, Gera, Germany. Permission granted to use under GFDL by Kurt Stueber, www.biolib.de
53, 55 From Jacob Sturm und Johann Georg Sturm, *Flora von Deutschland in Abbildungen nach der Natur mit Beschreibungen*, 1796, Nürnberg, Germany. Permission granted to use under GFDL by Kurt Stueber, www.biolib.de
57 Provided by Liaoning Provincial Museum
61 Photograph by Mark C. Fishman
68 Company archive Novartis AG
71 Pictorial Press Ltd / Alamy Stock Photo
82 Keystone / Science Photo Library / Sheila Terry
84 Copyright unknown
86 *An Experiment on a Bird in the Air Pump,* painting by Joseph Wright of Derby, © The National Gallery, London 2016
88 By permission of the Oesper Collections in the History of Chemistry, University of Cincinnati, Ohio
90 Wellcome Library, London
92, 94 Science & Society Picture Library
96 top Reproduced by permission, private collection
96 bottom Phillip A. Harrington

198 City of Cambridge, Massachusetts, 1877. Source: Norman B. Leventhal Map Center, Boston Public Library, http://maps.bpl.org/id/10184, Publisher: Franklin View Co.
202 Photograph by Novartis
249 Geography and Map Division, Library of Congress
252, 254 Photographs by MW Zander
255 Photographs by Wang Yi
350, 351 Photographs by Novartis
360 Photograph by Novartis

Iwan Baan
10–12, 17, 33, 38, 62–67, 114, 118, 122, 129, 131, 142, 148, 172, 179, 184–194, 205–208–212, 214, 215, 218, 220 bottom, 221, 224, 228–232, 240–244, 246, 256, 260–266, 270–286, 287 top, 288–304, 306–319, 320 top, 321, 326–333, 338–346, 352, 354

Andri Pol
14–16, 18, 19, 34–37, 58, 74–78, 101, 107–112, 115, 116, 119–121, 125–128, 130, 132–139, 144, 150–170, 174–178, 180–183, 213, 216, 220 top, 222, 233–239, 245, 289 bottom, 305, 320 bottom, 322, 323, 334, 355

FURTHER READING

Evolution

Browne, Janet. *Charles Darwin: The Power of Place*. New York: Knopf, 2002.

Darwin, Charles. *On the Origin of Species.* London: John Murray, 1859.

Gerhart, John, and Marc Kirschner. *Cells, Embryos, and Evolution*. Malden: Blackwell, 1997.

Nusslein-Volhard, Christiane. *Coming to Life: How Genes Drive Development*. San Diego: Kales Press, 2006.

Thompson, D'Arcy Wentworth. *On Growth and Form*, abridged and edited by John Tyler Bonner. Cambridge: Cambridge University Press, 1961.

The Garden and the Tar Pit

Daemmrich, Arthur, and Mary Ellen Bowden. "Top Pharmaceuticals: Introduction: Emergence of Pharmaceutical Science and Industry: 1870–1930," *Chemical & Engineering News* 83, no. 25 (June 2005).

Dale, Henry. "Paul Ehrlich," *British Medical Journal* 1.4863 (1954): 659–63.

Krikler, Dennis M. "Withering and the Foxglove: The Making of a Myth," *British Heart Journal* 54, no. 3 (1985): 256–57.

Qingxi, Lou. *Chinese Gardens*, translated by Zhang Lei and Yu Hong. Beijing: China International Press, 2003.

Martin R. Wilkins, Martin J. Kendall, and Owen L. Wade. "William Withering and Digitalis, 1785 to 1985," *British Medical Journal* 290.6461 (1985): 7–8.

The Ur Lab

Allen, Thomas J., and Günter W. Henn. *The Organization and Architecture of Innovation: Managing the Flow of Technology*. Amsterdam: Elsevier, 2007.

Cain, Susan. *Quiet: The Power of Introverts in a World That Can't Stop Talking*. New York: Crown Publishers, 2012.

Einstein, Albert. "Principles of Research" (1918). *The Collected Papers of Albert Einstein*, vol. 7, *The Berlin Years: Writings, 1918–1921*, translated by Alfred von Engel. Princeton: Princeton University Press, 2002.

Goldstein, Roger N. "Architectural Design and the Collaborative Research Environment," *Cell* 127, no. 2 (2006): 243–46.

Shenk, Joshua Wolf. "The End of 'Genius,'" *New York Times*, July 19, 2014, http://www.nytimes.com/2014/07/20/opinion/sunday/the-end-of-genius.html.

Can Buildings Drive Sociology?

Ede, Andrew, and Lesley B. Cormack. *A History of Science in Society: From Philosophy to Utility*. Peterborough: Broadview Press, 2004.

Holmyard, Eric J. *Alchemy*. Mineola: Dover Publications, 1990.

Leslie, Stuart W. "'A Different Kind of Beauty': Scientific and Architectural Style in I. M. Pei's Mesa Laboratory and Louis Kahn's Salk Institute," *Historical Studies in the Natural Sciences* 38, no. 2 (2008): 173–221.

Morris, Peter J. T. *The Matter Factory: A History of the Chemistry Laboratory*. London: Reaktion Books, 2015.

Shapin, Steven. *The Scientific Life: A Moral History of a Late Modern Vocation*. Chicago: University of Chicago Press, 2008.

Author, Mark C. Fishman, MD

Mark Fishman is a scientist and physician. He is currently
Professor in the Department of Stem Cell and Regenerative
Biology at Harvard. From 2002 to 2016 he was president
of the Novartis Institutes for BioMedical Research (NIBR),
the drug-discovery arm of Novartis Pharmaceuticals.
Prior to NIBR he was Chief of Cardiology and Director of
the Cardiovascular Research Center of the Massachusetts
General Hospital (MGH), where his research group helped
to introduce genetics in the zebrafish as an approach
to discover genes that direct embryonic development and
disease. He is an author of more than two hundred papers
and of the textbook *Medicine*. He has designed laboratories
at the MGH site in Charlestown, Massachusetts, and for the
new campuses for Novartis in Cambridge, Massachusetts;
Basel, Switzerland; and Shanghai, China.

ACKNOWLEDGMENTS

I thank Chris Klee, who was a true and patient partner
in bringing these projects to fruition, and in laughing
at the absurdities we encountered; Dan Vasella, who introduced me to architecture, provided me the opportunity to
undertake such a grand vision, and showed me how to tackle
the decisions needed; Jörg Reinhardt and Joe Jimenez,
who supported me throughout, especially in navigating the
late speed bumps; En Li and Amber Cai, whose work with
the teams in Shanghai quietly made sure I understood
the sometimes unexpressed constraints; Toshiko Mori and
Maya Lin, who fit the Cambridge campus into the complicated crossroads of Cambridge; Yungho Chang, who
brought the correct sensibility to the overall campus in
Shanghai; Tschekav Muench, who surmounted every moun-
tain and crossed every set of rapids to bring home the
Shanghai campus; Brian Lynch, whose quiet resolve meant
I could bring together the artistic with engineering talents
without any party exiting stage left; Jonathan Spector, for
input about alchemists; Vittorio Lampugnani, whose schol-
arly and experienced input gave me the comfort to make bold
choices while staying within boundaries; the photographers,
Andri Pol and Iwan Baan, who captured what we tried to do
and saw even more than we had imagined; Lars Müller, who
gave soul to the design of this book; and of course to all
the brilliant architects who brought such imagination and
personal emotion and clarity of vision to make all these
buildings and the campuses so special, while embracing
the clients' input, which they translated into proper
"architecturalese."

Mark C. Fishman

LAB
Building a Home
for Scientists

Concept: Mark C. Fishman
Editorial assistance: Lars Müller
Photographs: Iwan Baan, Andri Pol
Copyediting: Keonaona Peterson
Proofreading: Kristie Kachler
Image research: Muriel Blancho,
with a special thank you to Adam Xu, Shanghai
Design: Lars Müller
Design assistance: Anice Grossenbacher
Lithography: Print Professional, Jan Scheffler, Berlin, Germany
Printing and binding: Kösel, Altusried-Krugzell, Germany
Paper: LuxoArt Samt, 150 gsm

Lars Müller Publishers
Zürich, Switzerland
www.lars-mueller-publishers.com

ISBN 978-3-03778-497-6

Printed in Germany